Public Landscape Integration: Streetscape Facilities

公共景观集成
——街景设施

金盘地产传媒有限公司　策划
广州市唐艺文化传播有限公司　编著

中国林业出版社
China Forestry Publishing House

序言

随着公共景观设计在世界各城市的复苏以及人们日愈增强的意识，公共景观设计的重要性在政府的大力宣传以及支持下显得尤为突出。通过设计公共景观，城市环境得到提升，城市的身份也得以在公众面前建立，而且城市街道的形象会得到提高。注重功能性，安全性，美学功能的公共景观不仅可以促进一个城市的发展，还可以提高城市居民的生活水平。因此，公共景观设计并非只是针对一个人或是一圈人，而是适用于所有的人。

景观不仅存在公共环境中，而且建筑本身也是景观。随着人们对生活环境以及水平的要求日愈变高，人们更希望看到一些新鲜的独特的景观。本套书除了囊括国内外知名的公共景观设计，还包括各种独特的建筑景观。例如挪威 Ornesvingen 观景台，不仅拥有优美的建筑结构，而且将周围的环境映衬的更加壮丽，吸引了很多来自全世界的游客。

在城市中，街道起着决定性的作用。从当初的交通运输作用，现在的街道具有全面，强大的作用。而人们也希望穿梭于舒适，功能齐全，美观的街道。这样一来，街道设施，如：座椅，自行车停放架，候车亭，照明设施等等都有了新的角色去扮演。

公共景观设计要求公共景观和设计二者和谐统一。同时，公共空间中的设施也需要合理设计与安排。

本套书共有三册，每册书各有重点，特色。形象的图标带来更加直接，丰富的信息。第一册是"公共景观"，主要包括国内外知名的广场，公园和游乐场设计。第二册是"建筑景观"，主要包括以下三个类别：建筑造型和建筑结构，观景台，标识。创意的建筑造型和建筑结构，总能带给人新鲜的感觉。大胆的观景台设计，不仅吸引着游客，而且值得人们参考设计。极具功能作用的标识设计，形象而直接。第三册是"街道设施"，主要包括：座椅，自行车停放架，候车亭，照明设施，自动饮水器，垃圾桶和树木防护装置。这些街道设施不仅覆盖面广，而且具有强大的功能性，更加美化了城市街道。

这套书在一个更高的层次上来诠释良好公共环境的形成，可以提高人们的生活水平。特别是随着城市环境数字化的加强，舒服的公共空间设计和便利的公共设施让人们的生活变得越来越高效和充满活力。

Preface

With the revitalization of public environment design in cities around the world, as also the increased recognition of citizens, the importance of public design has emerged for the construction of a guideline which is maintained and managed by government integratively. Through the improvement for city environment by public design, the establishment of city's own identity and the creation of street image in a view of pedestrian is to get higher. A quality of a city is reflected by functionality, safety and aesthetics simultaneously. Therefore, the meaning of public environment design is not for some unique, specific or a minor group, but for several people to use or see for the creation of beautiful city landscape, which also means a design for all kind of people.

Landscape not only exists in the public environment, and the architecture itself can also be a landscape. With people's increasingly high requirement of living environment and living level, people prefer to see some fresh and unique landscape. In addition to including well-known public landscape designs, this book also includes a variety of unique architectural landscapes. For example, Norway Ornesvingen viewing platform not only has a beautiful architectural structure, but also makes the surrounding environment more spectacular, and attracts a lot of tourists from all over the world.

In a city, the street plays a decisive role. From the start, the street was just used for transportation. However, the street now has a comprehensive and powerful role. At the same time, people want to walk through a comfortable, functional and beautiful street. As a result, street furniture, such as bench, bicycle rack, bus shelter, lighting facilities, etc., have a new role to play.

The public landscape design requires public landscape and design to be of harmony and unity. At the same time, the facilities in the public space need reasonable designs and arrangements.

This book contains three volumes, each with unique focus and feature. The icon images bring more direct and rich information. Volume 1, named "Public Landscape", includes famous squares, parks and playgrounds design from all over the world. Volume 2 is "Architectural Landscape", mainly including the following three categories: Architectural Formative Arts and Structure, Observation Platform, Signage. Innovative architectural formative arts and structure always give people a fresh feeling. The bold observation platform design not only attracts tourists, but also is worth referencing. The signage is functional, image and direct. Volume 3, "Street Furniture", includes Bench, Bicycle Rack, Bus Shelters, Lighting, Drinking Fountain, Trash and Trees Protective Device. These street facilities not only cover a lot, and with powerful features, make the city streets more beautiful.

The purpose of this book is to design a book with applied examples of the amenity elements in the street furniture and to accomplish the public facilities on a higher level as one of the design method to upgrade the quality of citizens' lifestyle in the city environments. Especially with the stress of digital city environments development, comfortable public space and street furniture make citizens' lifestyle more energetic and effective.

目录 Content

第三册 街景设施
Vol.3 Street Furniture

 座椅
Bench ·· 010

材料景观展——灯光座椅
Material Landscapes —— Lighting & Bench

阿纳姆 Eusebiushof
Eusebiushof Arnhem

最长的座椅
The Longest Bench

Jonite 街具——座椅
Jonite Street Furniture —— Bench

BD Love 灯椅
BD Love Lamp Bench

Wanderest 人体工程学座椅
Wanderest Ergonomic Bench

Sebastien Wierinck 104 艺术中心座椅
Bench by Sebastien Wierinck for 104

支离破碎的果园——座椅
A Fragmented Orchard —— Bench

长椅
Long Chair

加高的街上座椅
Street Elevated Bench

手风琴家具
Morphing Furniture

Muscle 系列长椅
Muscle Bench

马瑟韦尔汉密尔顿路附近的座椅
Benches around Hamilton Road in Motherwell

座椅——David Shaw
Bench ——David Shaw

玻璃混凝土座椅
Glass Concrete Bench

Twig 系列混凝土座椅
Twig Concrete Bench

树墩座椅
Trunks by Malafor Benches

新型城市街具
New Urban Street Furniture

野餐用桌和板凳
Picnic Table & Bench

Mollymook 海滩街具
Mollymook Beach Street Furniture

Urban Amorfurniture 座椅
Urban Amorfurniture Bench

Marine 系列座椅
Marine Series

Twig 塑料灯座椅
Twig Plastic & Lighting Bench

"飞行甲板"座椅
Flight Deck Bench

Chelsea 城市座椅
Chelsea Bench

南岸公园座椅
South Bank Precinct Bench

Interference 座椅
Interference Urban Bench

布卡克沙滩海滨大道街具
Bulcock Beach Esplanade Street Furniture

Liana Lounge 街道座椅
Liana Lounge Street Bench

南岸科技学院街具系列
Southbank Institute Street Furniture

Springfield 湖边座椅
Springfield Lakes Bench

石制户外座椅
Stone Outdoor Bench

历水湾角阳光海岸座椅
Alexandra Headlands-Sunshine Coast Bench

Ondine 座椅
Ondine Bench

Longo 石基底座椅
Longo Stone Bases Bench

柔软座椅
Soft Bench

东京座椅
Tokyo Bench

公共座椅
Public Bench

Sucker Punce 座椅
Sucker Punce Bench

Tejo Remy 和 Rene Veenhuizen 个人作品展
——座椅
Tejo Remy and Rene Veenhuizen Solo Exhibition
—— Bench

S 形城市座椅
S Urban Bench

吸烟者座椅
Smoker Bench

自行车式街道座椅
Bike-Inspired Street Furniture

散步广场
Promenade Square

石头状的座椅
Stone Bench

BD Love 座椅
BD Love Bench

座椅和广告板两用系列
Bench & Billboard

Section2 系列座椅
Section 2 of Street Furniture

乡村公用座椅
Rural Township Bench

Dutton 公园座椅
Dutton Park Bench

意大利废弃铁轨改造项目
Retired Italian Railroad Transformed

Piano 系列
Piano Street Furniture

Toast 4 沙发
Toast 4

VIAS 空间
VIAS Space

拉斯内格拉斯滨水区座椅
Las Negras Waterfront Bench

意大利面条长椅
Huge Spahetti Bench

Mount Fuji Architects Studio 设计的座椅
Mount Fuji Architects Studio Bench

促进交际的另类会面碗
Offbeat Meeting Bowls Promote Community

Vekso 座椅
Vekso Bench

Rodeo 座椅
Rodeo Bench

北湖住宅区座椅
North Lakes Estates Bench

Zero Collection 座椅
Zero Collection Bench

信纸座椅
Letter Bench

跷跷板式座椅
Seesaw Bench

Energy-Moke 座椅
Energy-Moke Bench

Ensemble 座椅
Ensemble Bench

悬臂式座椅
Cantilevered Bench

Muungano 座椅
Muungano Bench

弧坑座椅
Crater Bench

Botanist 系列桌凳
Botanist Table and Bench

Hello Stranger 系列座椅
Hello Stranger Bench

Philly Pods 座椅
Philly Pods Bench

Dantelli 座椅
Dantelli Bench

Hammock 座椅
Hammock Bench

Wheels 座椅
Wheels Bench

Myriapoda 座椅
Myriapoda Bench

新型座椅
Street Space Nrgzers Bench

Shared Space Ⅲ 座椅
Shared Space Ⅲ Bench

Costco in Melbourne, Australia
科思科大楼景观设计

挤压座椅
Extrusions Bench

自行车停放架
Bike Rack ·············170

纽约市城市自行车停放架
City Bike Rack Design for New York City

多伦多自行车停放架
Toronto Bike Rack

港市自行车停放架
Bay City Bike Rack

绿色自行车停放架
Green Bike Rack

弗里蒙特自行车停放架
Fremont Bike Rack

Holbech 自行车停放架
The Holbech Bike Rack

三角形自行车停放架
Trio Bike Rack

B-Park
B-Park

国会大厦自行车停放架
Capitol Bike Rack

可丽耐座椅自行车停放架
Corian Bench Bike Rack

奥林匹亚自行车停放架
Olympia Bike Rack

"郁金香奇趣"和"草"自行车停放架
Tulip Fun Fun and Grazz Bike Racks

雕塑式自行车停放架
Sculptural Bike Rack

巨型梳子自行车停放架
Gigantic Comb Bike Rack

John Barbier 自行车停放架
John Barbier Bike Rack

小轿车式自行车停放架
Car Bike Rack

Ian Mahaffy 自行车停放架
Ian Mahaffy Bike Rack

候车亭
Bus Stop & Shelter ·············210

荷兰 Zuidtangent 专线上的候车亭
Bus Stop on Zuidtangent, Netherlands

巴西古里提巴候车亭
Bus Stop on Curitiba, Brasil

未来派联网候车亭
Futuristic Networked Bus Stop

未来派联网候车亭
Futuristic Networked Bus Stop

Cemusa 西班牙一线
Cemusa Spain Line 1

霍夫多普汽车站
Hoofddorp Bus Station

多伦多市汽车站
City of Toronto Bus Stop

La Dallman 汽车站
La Dallman Bus Stop

Sean Godsell 建筑事务所设计的汽车站
Bus Stop —— Sean Godsell Architects

安曼新候车亭
Amman's New Bus Shelter

候车亭
Bus Shelter

Laing O'Rourke 候车亭
Laing O'Rourke Bus Shelter

威灵顿市候车亭
Wellington City Bus Shelter

Euro Modul 候车亭
Euro Modul Bus Shelter

Montriol 候车亭
Montriol Bus Shelter

Adshel 广告展示候车亭
Adshel Advertising Display Bus Shelter

Coleman 候车亭
Coleman Bus Shelter

汉堡汽车站 / 候车亭
Bus Station & Shelter, Hamburg

The Lakes 街道设施
The Lakes Street Furniture

照明设施
Street Lighting ········· 256

海滩之灯
Licht am Strand

赫尔本博物馆的灯
Field of Light at the Holburne Museum

街头灯管
Street Lighttube

1000 封诗信
1000 Poems by Mail

字母路灯
The Street Alphabet Lamp

Garscube Link
Garscube Link

Corvin Gate 街道照明
Corvin Gate Public Light

3Rivers 行人交通灯
3Rivers Pedestrian Light

米兰 2010 年国际 LED 嘉年华
Milan LED Light Festival 2010

行人灯柱
Light Column Pedestrian

Quartier Des Spectacles 照明规划
Quartier Des Spectacles Lighting Plan

3XN 哥瑟姆路灯
3XN Gotham Street Lighting

B.Lux's 新品牌路灯
B.Lux's Brand New Street Light

骑士护柱
Knight Bollard

轻型护柱
LightScale Bollard

照明护柱
Light Column Bollard

霞关广场翻新
Kasumigaseki Plaza Renewal

K3 一期
K3 – Phase 1

Giulianova 纪念海滨
Lungomare Monumentale di Giulianova

自动饮水器 Drinking Fountain ······290

DRINKMi 自动饮水器设计
DRINKMi Drinking Fountain Design

GlobalTap 自动饮水器设计
GlobalTap Drinking Fountain Design

UBAN 自动饮水器设计
UBAN Drinking Fountain Design

阿波罗 400 自动饮水器设计
Apollo 400 Drinking Fountain

Hydro 300 自动饮水器设计
Hydro 300 Drinking Fountain

垃圾桶 Litter Bins ······298

花托式垃圾分离箱
Dispatch Receptacle Litter Bins

轨道形垃圾分离箱
Orbit Receptacle Litter Bins

三角形垃圾分离箱
Triad Receptacle Litter Bins

运输型垃圾桶
Transit Litter Bins

树木防护装置 Tree Protection ······306

尖沙咀树木保护
Tsim Sha Tsui Tree Protection

国家原住民卓越中心
National Centre of Indigenous Excellence

座椅
Bench

材料景观展——灯光座椅
Material Landscapes —— Lighting & Bench

项目档案

策划：Liane Hancock
项目地点：密苏里州圣路易斯市谢尔顿艺术廊

Project Facts

Curator: Liane Hancock
Location: Sheldon Art Gallery, St. Louis, Missouri

Material Landscapes 是最近在美国密苏里州圣路易斯市谢尔顿艺术廊展出的一项作品。该次会展的策划人是路易斯安娜理工大学助理教授 Liane Hancock。会展的主要内容汇聚了国际著名设计师的当代景观建筑。该次展出的项目多以照片和图纸的形式呈现，设计涉及的地址、规模和项目各有不同，但都体现了设计者在材料上的创新。其中最引入注目的是一个以图案点缀的花园，设计者创新地采用沥青铺地，并在公园里设计了一个外观美丽、意象深远的纪念馆。

Material Landscapes is an exhibition that recently opened at the Sheldon Art Gallery in St. Louis, Missouri. The show is curated by Liane Hancock, Assistant Professor at Louisiana Tech University. It features materiality in contemporary landscape architecture through projects by a group of national and international landscape architects. The projects shown in Material Landscapes exhibition are presented through photographs and drawings. The designs range in location, scale, and program, but they each address material conditions in an innovative manner. Highlights of the exhibit include designs using parametric modeling and common materials to create a park with a unique appreciation for asphalt, and a thoughtful beautiful memorial.

阿纳姆 Eusebiushof
Eusebiushof Arnhem

项目档案

景观设计：Strootman Landscape Architects
项目地点：荷兰，阿纳姆

Project Facts

Landscape Architecture: Strootman Landscape Architects
Location: Eusebiushof, Arnhem, The Netherlands

该办公楼位于阿纳姆Eusebiussingel，由Pi de Bruin和Royal Haskoning共同设计，是现有市政办公机构的延伸建筑。在地下停车场的地面上有三个庭院：Voorhof、Binnenhof和Expeditiehof。而受Eurocommerce Holding BV委托，设计师只设计了Voorhof和Binnenhof这两个庭院。

庭院属于时尚的抽象风格，其布局展示了一幅景观的图标。庭院的设计灵感一部分来源于城市周边的景观，另一部分来自其他景观。庭院布局既为来访贵宾呈现了清晰的辨识标志，又具有迷人的神秘色彩。抽象的巨大鹅卵石代表河岸，松树代表山区，绿色的小山丘代表连绵起伏的地面景观，蕨类代表森林，瓢虫代表阳光，白色围栏代表着牧马场，所有的景致都能引发游客的无限想象，带领他们去往理想中的胜地。Voorhof庭院被定义为"城市的门廊"，这里弥漫着一种特殊的氛围。白天，游客们穿过巍峨而又华丽的大门进入庭院内部，在这里小憩，欣赏周围的美景，也可以与朋友约在这里聚会。庭院内的路面以鹅卵石、灰绿色碎石和深灰色混凝土铺出美丽的图案。灰绿色的碎石与周围建筑以天然石板构建的幕墙在色调上保持浓淡一致，地面上的圆形斑点图案大小不一，是夜间的一大亮点。

A new offices complex has been built along Eusebiussingel in Arnhem which was designed by Pi de Bruin (Architekten Cie) and Royal Haskoning. The building is an extension to the existing municipal offices of Arnhem. At the surface level on top of the basement parking garage there are three courtyards: Voorhof, Binnenhof and Expeditiehof. On commission for Eurocommerce Holding BV, we have designed the layout of the Voorhof and the Binnenhof.

In an abstract fashion the layout of the courtyards refers to scenic icons, which are partially taken from the surroundings of Arnhem and partially from other landscapes. For the visitors, the layout also leads both to a recognition as well as to an estrangement and a smile. Abstracted giant pebbles refer to a river beach, pines refer to the Veluwe, a green hillside refers to undulating landscapes, ferns refer to the forest, ladybirds refer to sunny fields, and a "white-picket-fence" refers to horse ranches. The scenic references cause the visitors to conjure up thoughts about other places and offer an opportunity to distance oneself somewhat from the delusion of the day.

We perceive the Voorhof as an "urban lobby" a representative place with a special atmosphere, where you take a seat for a short while before going inside or where you make appointments with others. During the day the Voorhof is accessible to the public via a high portal and a fancy gate. The paving comprises a designed cobbled panel of dark grey concrete (2.00x1.30 m), which is filled with stone chippings that ensures a beautiful contrast of texture. The colour of the stone chippings is grey-green and has the same shade as the natural stone slabs on the facades of the surrounding building. Variously sized dome-spots in the flooring bring about a cloud of little highlights in the evenings.

座椅

最长的座椅
The Longest Bench

项目档案

设计：Studio Weave
项目管理和工料测量：Jackson Coles

Project Facts

Design: Studio Weave
Project Managers and Quantity Surveyors: Jackson Coles

2010年7月30日，该座椅在英国西萨塞克斯郡利特尔汉普顿街头公开展出。项目的设计者是 Studio Weave。这条可同时容纳300人的座椅沿着利特尔汉普顿人行道一直蔓延，不远处则是举世闻名的蓝旗海滩。流线形椅身沿着人行道蜿蜒前行，绕过灯柱，折到箱体建筑物的后方，然后骤然没入水面，形成沙滩与绿地之间的通道。它就像一条沿海滨的人行散步道，静静地躺在项目的所在地，然后不着痕迹地渐渐融入周边的环境；它又好似一个外观迷人的手镯，将步行道串联成一个整体，集中展示沿途特色的景点。沿着长椅分布的还有两个铜制的环形建筑承接绿地和步行道，当长椅蔓延到封闭的环体内部时，原本井然有序的结构突然变得混乱起来，它们时而跳到墙壁上，在墙壁的表面形成座椅，时而蹦到天花板上，变成顶部的天窗。环体内部汇集各种奇形怪状的座椅，座椅本身亦是一个囊括周边美景的巨型镜框。

The longest bench in Britain was opened to the public in Littlehampton, West Sussex on the 30th July 2010. The bench seats over 300 people along Littlehampton's promenade, overlooking the town's award-winning Blue Flag beach. Designed by Studio Weave, the structure sinuously travels along the promenade, meandering around lampposts, bending behind bins, and ducking down into the ground to allow access between the beach and the Green like a seaside boardwalk. The Longest Bench rests gently on its habitat and adapts to its surroundings like a charm bracelet, which connects and defines the promenade as a whole, underlining it as a collection of special places that can be added to throughout its lifetime. Accompanying the long bench are two bronze-finished steel monocoque loops that connect the promenade with the green behind it. As the bench arrives inside the twisting loops, it goes a little bit haywire, bouncing of the walls and ceiling creating seats and openings. The loop contains the haywire stretch of bench and frames the views each way.

Bench

Jonite 街具——座椅
Jonite Street Furniture —— Bench

Jonite 街具采用一种新型的令人兴奋的街道装饰材料,这种新的理念将从本质上颠覆人们对石材的陈旧观念。Jonite 石材不仅绿色环保,它最大的特色在于如果设计方法正确,材质本身就会变得特别轻。与天然石材相比,Jonite 石材具有极高的机械完整性,可最大限度地延长材料的使用寿命,还不易褪色。

Jonite is now launching Jonite Street Furniture, an exciting new concept of street furniture. This new concept is attempting to eliminate the perception of stone as a bulky and heavy material. Jonite stone material is not only Green Label certified, but also significantly lighter when designed correctly. It has also extremely high mechanical integrity which greatly increases the duration it takes for the material to age and fade as compared to natural stones.

座椅

BD Love 灯椅
BD Love Lamp Bench

项目档案   Project Facts

设计：Ross Lovegrove   Design: Ross Lovegrove
项目地点：巴塞罗那   Location: Barcelona
完成时间：2010   Year: 2010

如果说由 Ross Lovegrove 设计的 BD Love 灯椅系列是一项根本性的创新，那么这个座椅和灯饰混合物将在室内外公共空间的装饰领域里开辟出一片新的天地。Ross Lovegrove 倾注所有情感，将照明灯安装在雕塑般生动的公共座椅上。BD Love 灯椅系列可同时供四个人坐下，孩子们也可以在上面玩个痛快。
座椅上方灯具的制作材料为旋转模制聚乙烯，并且提供多种颜色选择（包括最华丽的荧光红），内部既可以填充细沙，也可以用水代替。该座椅符合 IP65 等级，适用于户外操作。

If the BD Love Lamp Bench designed by Ross Lovegrove for this programme was a radical innovation, this hybrid between a seat and a lamp opens up a new territory in furnishing concepts for internal and external public areas. Ross Lovegrove has created public seating with a lamp that has all the exuberance of lively sculptural forms, designed to engage the emotions. Its generous proportions enable it to seat up to four people at once and the children to have a whale of a time!
The lamp is made from rotation moulded polyethylene and is available in a host of colours (including a spectacular fluorescent red) and can be filled with either water or sand. The bench meets the standard of IP65 rated for outdoor use.

Bench

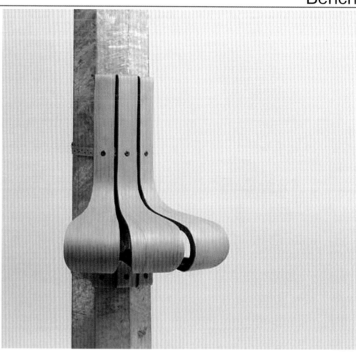
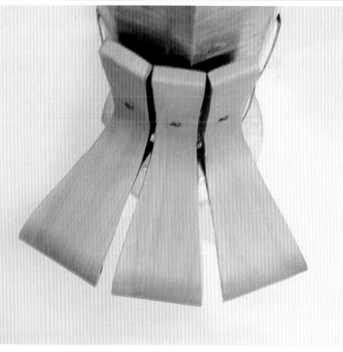

 Wanderest 人体工程学座椅
Wanderest Ergonomic Bench

设计：Nichola Trudgen   Design: Nichola Trudgen

让老年人保持身体健康和活力的最佳方法是做一下短距离的散步。尽管短距离散步的运动量较少，但是对于老年人来说，可以将散步的距离分成几个可操作阶段，中途作短暂的休息，同时欣赏周边的景色。不幸的是，位于养老院和退休村的公共座椅通常设置得不够。为了解决这一问题，Wanderest 这一种结构科学、设计符合人体工程学的座椅便诞生了。三块同一规格的面板被固定在灯柱上，形成一个简易的座椅。为了将面板固定在一起，设计将一根不锈钢带固定在面板背面的凹槽里，然后用螺栓将钢带在两个部位固定。面板组合工作完成后，就可以将钢带和面板直接用螺栓固定到灯柱上。面板在人体接触的部位略微倾斜，高度适中，因此老人们无需弯腰便能舒服地坐下或起身。

The easy way for elderly to keep healthy and mobile is to do a short walk. Although a short walk can be considered as light exercise, the elderly need to break their walk into manageable distances, take some rest for small intervals, and enjoy the surroundings. Unfortunately, public seating is not placed frequently enough around rest homes and retirement villages. To accommodate these situations, Wanderest, a well engineered and ergonomic seat has been designed. It is constructed of 3 identical panels (extra panels can be added for a larger seat) that can be attached to a lamp post as a resting point. To combine the panels together, a stainless steel strap is put through the indentation at the back and the strap is bolted onto the panels at two points. Then the stainless steel strap can be bolted to the lamp post or attached with a clamp mechanism. The panels are shaped with a slight downward slope on the sitting part and installed at a perching height. By this way, the elderly can slide on and off the seat comfortably, there's no need to bend down just like conventional seat.

座椅

Sebastien Wierinck 104 艺术中心座椅
Bench by Sebastien Wierinck for 104

项目档案   Project Facts

设计：Lepold Lambert   Design: Lepold Lambert
项目地点：德国，柏林   Location: Berlin, Germany

Funambulist 是由 Lepold Lambert 设计的日用表演舞台。
项目的名称来源于两个启示：一个是设计使用的钢丝，它将整个背景一分为二，并牢牢把握住建筑的出入口，如此一来，在钢丝上的行走就变得轻松自如多了；另一个指代的是 1974 年 Philippe Petit 在世界中心两塔间表演高空钢丝行走这件事情。
比利时设计师 Sebastien Wierinck 运用大量结构灵活的聚乙烯管设计出一个雕塑般的装饰。从公共座椅到咖啡厅坐席以及临时座位，Sebastien Wierinck 的每件作品都具有令人震撼的视觉效果，并不断与环境发生互动。
盘丝状装饰的运用在现代室内并不常见，但由比利时设计师 Sebastien Wierinck 新近完成的创意长椅却赢得了众多的关注。触手一般的座位被安置在空旷的室内，展现了设计在物质创建和空间制作方面的才能。创意长板凳全部采用塑料管打造，同时还运用了计算机辅助制造技术。

The Funambulist is a daily architectural platform edited by Lepold Lambert based on the archives of the former boiteaoutils.
Its name is inspired by a reflection on the line as the architect's medium. In fact, this line on the white page that ends up spliting two milieus from one another, controls the access of the bodies. The act of walking on the line (funambulist is another word for tight-rope walker) thus becomes an act of freedom. It also refers to Philippe Petit crossing illegally the space between the two towers of the World Trade Center in 1974.

Bench

Belgian Design Sebastien Wierinck has been creating his installations of sculptural furniture made of flexible polyethylene tubes. From public benches to cafe seating to temporary installations, his pieces always challenge the way people view and interact with environmental space.

It is not often that scrolls of this architecture blog are taken over by modern interiors. However, the recent sculptural works of Belgian Design Sebastien Wierinck commanded such attention. The tentacled installations housed in otherwise blah interiors, exemplify a talent in both object creation and space making. Constructed of plastic tubing, the installations are formed using CAM technology.

支离破碎的果园——座椅
A Fragmented Orchard —— Bench

项目档案　　　　　　Project Facts

设计：OKRA　　　　　　Design: OKRA
项目地点：日本，东京　　Location: Tokoy, Japan
完成时间：2010　　　　　Year: 2010

OKRA景观事务所在城市公共空间设计的竞赛中获胜。由于该公共空间所在的地点缺乏一个清晰的项目，OKRA认为当务之急是为这个支离破碎的环境创建一个统一的身份，因此他提出在该地打造团队的个人项目，即都市果园。通过栽种苹果树和梨树，OKRA希望可以在分散的空间和建筑之间形成一种凝聚力，而有机食品城市的打造恰巧赋予了公共空间这一属性。都市果园由许许多多的小果园组成，内部氛围各有千秋，园内的水果可任意采摘。

所有的小果园以块状形式分散布局，果园间的道路结合公共建筑和私人物业呈网格系统分布。果树类型的变换和种植密度的更新在高层商业和城市建筑之间形成有效的过渡，同时使南部的建筑与旁边的Klosengroendchen公园有机地融合在一起。同时，项目还运用了一系列线条结合各个空间。

OKRA won the competition to design a public space between office buildings in the Gruenewald area of the Luxemburg city borough of Kirchberg. OKRA thought of how to give the fragmented neighbourhood a consistent identity of its own. As the area lacked a clear programme, OKRA chose to come up with their personal programme for an "urban orchard" By planting apple and pear trees, OKRA has attempted to create cohesion between the scattered spaces and buildings. Thus the theme of the "edible city" gives this public space its coherence. The urban orchard consists of many little orchards, each with its own atmosphere. Passers-by are invited to pick the fruit.

Bench

The orchard stands in a fragmented layout of squares and passages organized in a grid system together with buildings and private parcels. A gradual change in the types and the densities of plantation create a high quality link between the higher part of the district, more business and urban alike and the southern part, more natural, leading towards the Klosengroendchen park. A network of lines bind the different pieces of public space in order to create a narrative and spatial link between each of the existing public spaces fragments.

座椅

长椅
Long Chair

项目档案	Project Facts
设计：Jo Nagasaka	Design: Jo Nagasaka
项目地点：加拿大，多伦多	Location: Toronto, Canada
完成时间：2010	Year: 2010

Long Chair 是由一连串钢管组合成的长椅。首先，将钢管弯成座椅的造型，取用截面形状。其次，不断重复第一步骤，直到截面结构达到长椅所需的尺度。完成后的长椅全部捐赠给日本工业大学世纪纪念大楼。所有的长椅均是由无缝钢管打造的。

Long Chair is a bench made from a continuous steel tube, designed by Jo Nagasaka of Schemata Architecture Office. The tube is bent into a chair profile which is repeated along the length of the bench. We designed a long chair and donated it to celebrate the completion of Centurial Memorial Building of Nippon Institute of Technology, says Nagasaka. This chair is made of one seamless steel pipe which draws the section shape of chair. Other steel pipes also draw various ways to sit.

Bench

座椅

加高的街上座椅
Street Elevated Bench

项目档案　　　　Project Facts

设计：Itay Ohaly　　Design: Itay Ohaly
项目地点：以色列　　Location：Israel
完成时间：2011　　Year：2011

这些加高的座椅依附在街上的灯柱上，成螺旋形梯子状。人们可以在座椅上欣赏城市的风景，也可以和朋友窃窃私语。

The "Elevated Bench" is urban furniture with a view. The public benches are constructed at the top of spiral staircases that climb streetlight poles, providing a new perspective on the city streets, as well as a more private space for conversation.

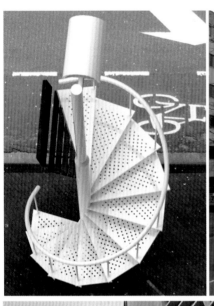

Bench

手风琴家具
Morphing Furniture

项目档案　　　　　　　Project Facts

设计：Noiz Architects　　Design: Noiz Architects
项目地点：以色列　　　　Location：Israel
完成时间：2011　　　　　Year：2011

手风琴家具将完整的立体造型简化成了不连续的切片，看上去就像是手风琴一般，似乎轻轻的一个移动就能让它弹奏出华丽的乐章。现有风琴椅和风琴桌两种款式，从某些角度看去，曲线如群山起伏，相当华丽。

This kind of furniture simplifies the complete three-dimensional shape into uncontinuous slices, and looks like an accordion, a gentle move of which can play a gorgeous piece of music. There are two styles: bench and table, which are quite gorgeous like curing hills seen from a certain angle.

座椅

 Muscle 系列长椅
Muscle Bench

项目档案	Project Facts
设计：Lexandre Moronnoz	Design: Lexandre Moronnoz
项目地点：法国	Location: France
完成时间：2011	Year: 2011

只需看上一眼就能想象出它的美妙，这便是 Muscle 系列长椅。Muscle 系列长椅造型动感，线条纯粹，它从本质上否定了僵硬呆板的传统街道设施，极大程度地肯定了现代建筑结构和景观设计。椅面高低起伏不一，或坐或躺都是一种乐趣。像肌肉的纤维组织一样，设计将切割金属板压缩和拉伸维持椅面的刚度。
光滑的炮铜灰环氧防腐漆具有反光的作用，能够增加座椅整体的亮度。阳光透过钢材之间的缝隙洒下来，影子和实体交织在一起，整个座椅显得栩栩如生。

As spectacular to gaze at as it is comfortable to rest on. Muscle is a counterproposal to traditional stiff and motionless street furniture thanks to its dynamic forms and its purified lines. A piece compliments both contemporary architectural structures and modern landscape designs.
The bench offers the possibility of sitting or lying down in response to the surface's relief. Like the fibrous structure of a muscle, the cut metal sheets work with compression and tension to maintain the rigidity of the resting platform.
The smooth gun metal gray epoxy finish reflects the light and adds to the sense of lightness. As the sun passes through the metal bars, the shadows mix with the actual piece adding yet another dimension to this exceptional urban bench.

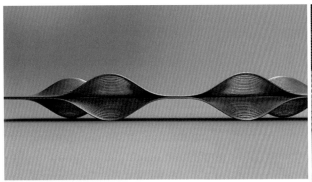

Bench

 马瑟韦尔汉密尔顿路附近的座椅
Benches around Hamilton Road in Motherwell

位于旧城市政厅和图书馆附近的马瑟韦尔汉密尔顿路是 Hardscape 特色长板凳的故乡。长板凳的材料为 Kobra 花岗岩、特殊的环氧树脂、木板和不锈钢板材，每一条都从自身的角度将城市的历史娓娓道来。

Hamilton Road in Motherwell around the Old Townhall and Library is home to some unique bespoke benches supplied by Hardscape. Each bench, produced in Kobra granite with individual white epoxy resin text and timber and stainless steel tops, tells its own story of some of the town's historical past.

座椅——David Shaw
Bench ——David Shaw

设计：David Shaw　　Design: David Shaw
项目地点：美国　　　Location: USA

本案是一个具有多种用途且非常有力的解决方案。与时俱进的造型与对材料的精挑细选造就了这些能够适用于各种街道环境的座椅。

This suite of furniture is a versatile and robust solution. The contemporary nature of the design forms and selection of materials allow it to be used in a variety of streetscape environments.

玻璃混凝土长椅
Glass Concrete Bench

项目档案　　　　　Project Facts

设计：Ivanka Design　　Design：Ivanka Design
项目地点：匈牙利　　　Location：Hungary

在设计长椅的时候，充分考虑到周围的环境，强调突出整个空间。壳体结构的长椅是由混凝土以及嵌入式玻璃构成的。

The environmental features were taken into account in the art-design planning of the bench which seeks to give emphasis to the space. The shell structure comprises of concrete with embeded glass elements.

座椅

Twig 系列混凝土座椅
Twig Concrete Bench

Twig 系列水泥凳在这个城市里随处可见，它们是城市旅行者驻足的空间。Twig 的外形设计参照了原生态的树枝打造，但混凝土实体又掩饰了设计的意图。Twig 系列水泥凳拥有圆润的边缘和令人着迷的模块化设计，是都市装饰的最佳选择。

Suited for a very modern area, the Twig concrete bench offers resting space for the urban traveler. The design resembles the form of a twig but the concrete masks any idea of eco-friendly intention. Twig is the perfect touch in a modern urban area. This system of benches invite to rest and enchant with its round edges and modular design.

树墩座椅
Trunks by Malafor Benches

该座椅由 Malafor 设计，是一款趣味性十足的产品。设计将一段结实的橡木用不锈钢铁皮包裹起来，创造了一个一端带着把手、表面光滑的不锈钢凳子，凳子还可以根据喜好喷上不同的颜色。

这把用横向截断的橡木构成主体，外层包覆不锈钢铁皮的座椅既有着镜子闪亮的外观，又有油画的罩光外层。除此之外，这种座椅还设计了把手，方便随身携带。

The Polish designer Malafor has designed a seat made of a solid piece of oak wrapped in steel. The stools have a metal handle on one side and are made with a polished stainless steel surface or painted in bright colours.

It is a seat made of a piece of oak wood cut crosswise and dressed in stainless steel either with a shiny, mirror-like, or painting finish. The trunk has a handle, so you can take it whenever you like without making any fuss about it.

座椅

 新型城市街具
New Urban Street Furniture

设计：Rocker Lange Architects

Design: Rocker Lange Architects

由 Rocker Lange Architects 建筑设计公司提供的设计方案以现代化的都市座椅体现了公司将街具纳入城市全盘设计所做的努力。在本案中，设计师尝试着用丰富多样的设计取代单调，缺乏活力的设计，并在创新的设计中加入多种元素，最终达到特殊的，健康的标准。本案还引入了香港的城市体验，并努力将设计与城市整体设计融为一体。云朵形状的座椅为城市营造了一种特殊的体验，这是城市规划中较为常见的现象，既孕育了城市的旅游业，又提升了市民的个人荣誉感。虽然这只能算是一种解决方案，但这也是大多数城市正需要的。

This design proposal from Rocker Lange Architects for a contemporary city bench seeks to understand the concept of street furniture as a holistic design problem. Instead of offering only one single static design, this scheme suggests multiple varying solutions that meet specific fitness criteria. As part of their submission, they are taking the "experience" of Hong Kong and really trying to relate design as part of a bigger picture such as how this could foster a unique brand experience in Hong Kong. A criteria is to be considered more in urban projects to build and nurture tourism and personal pride. Some would call it just a solution, and that is where the problem lies in many cities.

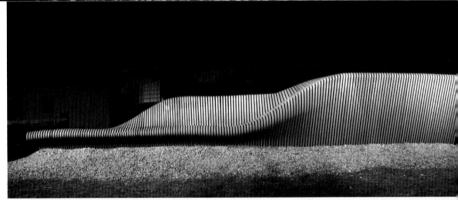

座椅

野餐用桌和板凳
Picnic Table & Bench

设计：Michael Beitz，美国

Design: Michael Beitz, USA

美国奥马哈市贝米斯当代艺术中心的围栏上挂着一张野餐用桌。该类餐桌是 Michael Beitz 的新品，材料为白杨木复合板、抗腐蚀环氧树脂和定制的模具。该类餐桌具有很强的实用性且对外开放，可同时容纳 10 人用餐，是贝米斯花园展和设计艺术廊的永久性辅助设施，同时也是将美国艺术家的街具概念方案付诸实际的首例。

Lanks of a picnic table are draped over the railing at the Bemis Center for Contemporary Arts in Omaha. Created by Michael Beitz, the table is shaped by laminating sheets of poplar with marine epoxy over a custom-made mould. The table is fully functional and open to the public, seating up to 10 people. The permanent installation is a project in conjunction with the Bemis Gardens exhibition and design gallery. This project is the first of the American artist's conceptual furniture drawings to be realised.

Bench

Mollymook 海滩街具
Mollymook Beach Street Furniture

设计：Miranda Lockhart   Design: Miranda Lockhart

项目的沿岸地理位置与功能需求是决定本案的关键因素。Mollymook 海滩街具使用材料包括切割钢材和定形板材，产品形式包括台式长椅、长桌和垃圾箱。
Mollymook 的设计可随着项目周边的景色、微风和日照条件时刻改变自身的外观。不锈钢设计上的图案外观是模仿了海浪在沙滩上留下的形状。从不同的角度看去，椅背的花纹与光影相互嬉戏，形成别样的视觉效果。

Aspects of the coastal location along with functional considerations became the key elements used to develop the Mollymook designs. Combining laser-cut steel with shaped hardwood timber elements, the furniture suite includes platform bench seats, table settings and bin enclosures. Taking advantage of the views, breezes and the sun conditions on site, the appearance of the furniture changes throughout the day. The patterning details in the steel elements of the furniture resemble the forms left by wave action in the sands. This effect is emphasised by the back to back arrangement on the platform bench, which plays with light and shadow effects that change according to the viewers' position.

 # Urban Amorfurniture 座椅
Urban Amorfurniture Bench

设计：塞尔比亚诺维萨德 Student Group a7

Design: Student Group a7, novi sad SERBIA

本案是根据人体工程学设计的，不仅体现了空间感，而且非常实用。Urban Amorfurniture 可同时提供 5 个人的空间，无论是从选材、后期制作以及最后的安装，该座椅的设计自始至终都贯彻了可持续发展的理念。厚度为 0.5 厘米的硬纸板用纯手工的方法进行多层压缩，然后塑造成立方体的几何形态。从外表看来它只是个不规则的立方体，内部却异常温暖。采用硬纸板塑形是非常不错的选择，手感也不错，最重要的是，硬纸板可以无限循环使用。

Project presents both space installation and utilitarian object (part of furniture), designed and carried out with anthropological measures with 5 seating places. Concept of sustainable development is consistently implemented throughout all phases of project, from selection of materials for processing to final construction. Cardboard plates are used, thickness 0.5cm, hand made and assembled in layers. These layers form structure which is geometrically shaped looked from outside (part of cube), and amorf inside. Cardboard as material is suitable for processing, friendly for touch and renewable (recycling process can be repeatedmany times).

座椅

Marine 系列座椅
Marine Series

设计：昆兰岛黄金海岸 Alexander Lotersztain

Design: Alexander Lotersztain, Queensland's Gold Coast

Sanctuary Cove 位于昆兰岛黄金海岸的度假社区。本案中的定制街具是专为一个名叫 Mariners Village（水手村）的地点而设计的。这里不仅是来往船只停泊的码头，同时也是休闲度假的中心，有特产店、时尚精品店、艺术画廊、餐馆等，各种商业琳琅满目。装配式的折叠钢板和可丽耐大理石结合在一起，既保证了结构的牢固性，同时又体现了 Marine 系列低调的美学。Marine 的设计灵感源于船只和帆船的造型，而细节设计则直接参照甲板和系缆桩。该系列街具多呈现为座椅、垃圾箱和单车架，既给用户带来了操作上的便利，又方便了后期的维修工作。

本案的选址分析、产品的可行性研究、原模和技术文件均由 Street & Garden 工作室提供。

Sanctuary Cove is a resort-style master-planned community located on Queensland's Gold Coast. This custom furniture suite was developed specifically for the Mariners Village, which features a marina catering for a variety of vessels. The Village is a central area of the resort for speciality shops, fashion boutiques, art galleries, and restaurants.
Fabricated from folded steel combined with white Corian, the Marine series of furniture is robust while maintaining a visually minimal and light design aesthetic. Inspired by the forms of boating and sailing vessels, the furniture makes direct reference to the details of ships decks and boat moorings.

Bench

The range included a variety of seating elements, bins enclosures and bicycle racks that met both accessibility and maintenance requirements of users.
All necessary site analysis, product feasibility studies, prototyping and technical documentation were also undertaken by Street & Garden Studio during the various design and manufacturing phases of the project.

Twig 塑料灯座椅
Twig Plastic & Lighting Bench

设计：澳大利亚 Street & Garden 工作室

Design: Street & Garden, Australia

Twig 塑料灯长椅设计得时尚动人，非常适用于装点现代化区域，又为城市游客提供了休息场所。其外形仿如小树枝，而塑料覆盖层的设计又透露这款长椅的环保特点。模块化的设计，圆滑的边缘邀请游客享受休闲的时光。

Suited for a very modern area, the Twig plastic & lighting bench seating offers resting space for the urban traveler. The design resembles the form of a twig but the plastic masks any idea of eco-friendly intention. Twig is the perfect touch in a modern urban area. This system of benches invite people to rest and enchant with its round edges and modular design.

Bench

041

座椅

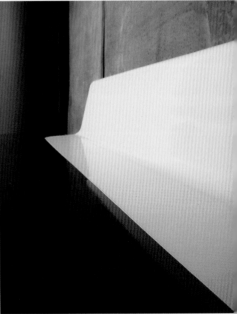

 "飞行甲板"座椅
Flight Deck Bench　　　　设计：Surya Graf　　Design: Surya Graf

由于本案的初衷是为布里斯班国际机场附近的一家酒店设计配用的座椅，因此所有的灵感都源于航空飞机的造型及航空材料。本案中的"飞行甲板"系列座椅采用可折叠的低碳钢组装，结构十分牢固，并在设计中加入了可延长使用寿命的元素。该座椅的背靠设计可根据需要随时进行调整，也可以按照用户的喜好更换色彩。
本案的选址分析、产品的可行性研究、原模和技术文件均由 Street & Garden 工作室提供。

Initially designed for a hotel project near the Brisbane International Airport, the Flight Deck Bench series was inspired by aeronautical forms and materials. Fabricated from folded mild steel, the pieces are robust and have been specifically designed for longevity in urban environments. The component backrest design allows for adaptability and colour variation. All necessary site analysis, product feasibility studies, prototyping and technical documentation were also undertaken by Street & Garden studio during the various design and manufacturing phases of the project.

Chelsea 城市座椅
Chelsea Bench

设计：纽约 Gnacio Ciocchini   Design: Gnacio Ciocchini, New York

Chelsea 城市座椅的灵感源于纽约动感的城市节奏、行人的来去匆匆以及城市的车水马龙。同时，Chelsea 灵感也来自于纽约市内用钢铁和玻璃搭建的公车候车亭、数字世界以及摩登的建筑。该类座椅经过工效学测试，充足的面宽能够容纳任何体型的使用者，而且坐上去非常舒适。同时，充足的面宽也可以使使用者之间保持适当的距离，多出的空间也可以用来放一些私人物品。早在调查阶段设计小组听到的最频繁的两种抱怨就是：椅子太窄，人与人靠得太近，一不小心就碰到对方，而且也没有地方放置包包这一类个人物品。椅子底座的弧度太大，看起来好像不太结实。
Chelsea 的主体部分包括一个 U 形不锈钢底座、一个轻质的镂空装饰、一个激光切割的碳钢板椅面和一个炭灰色电镀膜装饰。此外焊接的靠背也是镂空的。早在 2010 年初该座椅就已被纽约交通部和纽约公共设计委员会认定为城市的标准座椅。

The Chelsea urban bench design was inspired by the constant movement and urban rhythms of New York City, the fast dance of pedestrians and cars that are always on a rush to get somewhere. Other inspirations were the new New York stainless steel-and-glass bus shelters, the digital world, and contemporary architecture. The seats are comfortable——the ergonomics were tested with prototypes, and have a generous width to accommodate a wide range of the population. This feature makes sure users are not too close to each other while seating and creates space for personal items to be resting next to them. These were two complaints about urban benches the team heard several times during the research phase: insuficient seat width caused users to be too close to each other or touching, and there was no space for personal items, like handbags, to rest on the same personal seat space. The stainless steel arch base is very strong while giving the bench a light and floating appearance.
The main parts of the bench are a fabricated U-channel stainless steel base with a light bead-blasted finish, laser-cut and press-formed carbon-steel plate seats with a dark charcoal-grey powder-coated finish; and fabricated stainless steel plate armrests with a light bead-blasted finish. The product was approved in early 2010 by the New York City Department of Transportation and the New York Public Design Commission to be a standard bench for the City of New York.

座椅

 南岸公园座椅
South Bank Precinct Bench

设计：Surya Graf   Design: Surya Graf

本案是专为QPAC场地外停车站设计的特征元素，共同协作的设计公司还有Cox Rayner Architects。由于该地的交通流量非常大，因此急需一个结构牢固且容易维护的解决方案。考虑到座椅使用者的人数可能较多，座椅的长度达6米，整个座椅用不锈钢焊接的方式保证座椅的韧度和使用寿命，壁装式的设计也避免了座椅垃圾截留。除此之外，椅面还设计了防滑的结构，这也是该座椅的关键细节之一。

Developed in conjunction with Cox Rayner Architects, this seating solution was designed as a feature element for a bus stop outside the QPAC site. Catering for large groups of people, the seat is one with continuous 6m length. Being a high traffic area, the site required a very robust and easily maintained solution. Fabricated entirely from folded steel for strength and longevity, the wall mounted installation also helps to limit the possibility for rubbish entrapment. One of the key details in this piece was the integration of skate deterrents into the laser cut surface patterning.

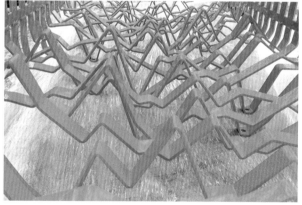

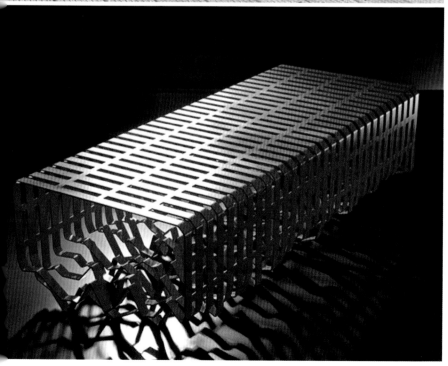

 Interference 座椅
Interference Urban Bench

设计：法国 Alexandre Moronnoz

Design: Alexandre Moronnoz, France

这是一个外表冷酷而内心火热的设计。它的结构无比清晰，内部构造十分动感，而且毫无掩饰，它是秩序与混乱之间的平衡，动静皆宜。不管是喧嚣的市区，还是秋千草地，Interference总是能够吸引人们的注意力。

An urban bench with a calm and organic exterior and a frenetic, moving, exposed interior. A balance between order and disorder, movement and stability. Equally at home amidst urban bustle or swaying grasses, Interference is an eye catching addition to the landscape.

布卡克沙滩海滨大道街具
Bulcock Beach Esplanade Street Furniture

设计：Surya Graf　　　　Design：Surya Graf

本案是专为布卡克沙滩设计的方案。该方案的灵感源于海岸主题，其主要目的是鼓励设置更多的公用设施。针对项目的可行性、耐用度和容易维修等要求，设计采用了激光切割折叠并经过阳极氧化的铝材、不锈钢材和数控加工的硬木板材等材料。项目的产品范围涵盖长椅、户外餐桌、吧台、系缆桩、单车架、沙滩淋浴设备、以及大量雕塑般的标志性设计。

Developed in conjunction with Place Design Group for the Sunshine Coast Regional Council, this extensive custom furniture suite was designed specifically for the Bulcock Beach Esplanade site. With forms inspired by coastal themes, the main focus of the furniture suite was to encourage the communal use of the site. Motivated by accessibility, longevity and ease of maintenance, the material palette for the suite included laser-cut, folded aluminium with anodised surface treatment, stainless steel, and CNC machined hardwood timber. This extensive suite includes a variety of bench seats, table settings, bar settings, bollards, bicycle racks, beach showers and a number of sculptural and signage elements.

Bench

座椅

  Liana Lounge 街道座椅
Liana Lounge Street Bench

设计：S & G / GMG   Design: S & G / GMG

Liana 系列长椅外观颀长而苗条，似一根木质的藤蔓牢牢地扎根地下，凭借周边的结构支撑自身。因此，在本案中，高密度聚乙烯包层钢被固定到预先做好的水泥主体上。绿色的高密度聚乙烯藤状靠背沿着水泥主体蜿蜒前行，形成大排的空位，鼓励行人上前探索和嬉戏。此外，椅身的设计预先加入了水分蒸发和夜间照明灯等元素。

The Liana is a long-stemmed woody vine which is rooted at ground level relying on surrounding structures for support. Following this, the Liana Lounge concept binds a sinuous HDPE clad steel structure to a pre-cast concrete host. The green HDPE vine winds its way across the concrete, creating a vast array of resting positions, encouraging exploration and play The pre-casting includes water shedding details and accommodates LED lighting transforming the structure by night.

Bench

座椅

南岸科技学院街具系列
Southbank Institute Street Furniture

设计：Alexander Lotersztain   Design: Alexander Lotersztain

SIT 系列完全采用不锈钢折叠而成，结构牢固，外观简洁轻巧。此外，设计师在颜色的选取上参考了学院的品牌形象，并采用激光切割的图案指代学校执教的学科。座椅、喷泉、桌子、工作台、树池、系缆桩、单车架和垃圾箱等设施，不仅操作简单，而且容易维修。

Fabricated entirely from folded steel, the SIT range is extremely robust while maintaining a visually minimal and light design aesthetic. The colour selection for this project makes direct reference to the branding identity of the Southbank Institute with the laser-cut pixel pattern representing the different areas of technology that are taught at the site. The range includes a variety of seating elements, drinking fountains, table settings, workstations, tree grates, bollards, bicycle racks and bins enclosures that met both user accessibility and maintenance requirements.

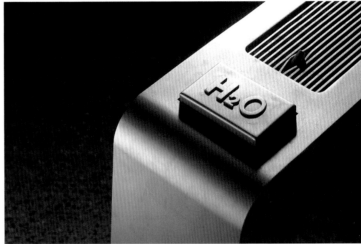

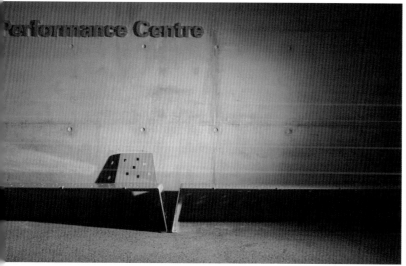

Springfield 湖边座椅
Springfield Lakes Bench

设计：Bjorn Rust　　Design: Bjorn Rust

本案是专为 Robelle Domain Parkland 定制的街道设施，是 Springfield 湖边住宅项目的一部分。为了控制成本、效率，同时寻求高品质的解决方案，设计将其重心全部放在找出使产品的使用寿命更长，更容易维修的解决方案上。

该套产品全部采用不锈钢折叠而成，材料在生产过程中得到了充分利用。激光切割的图案与建筑细节紧密地结合在一起，既能防滑又能够加速表面的水分蒸发。该套产品包括无靠背长椅和户外长桌，其中采用模块化设计打造的长桌可以配合场地的需要创造出不同的外形。

This custom furniture suite was designed specifically for the Robelle Domain parkland, as part of the Springfield Lakes residential development. Motivated by the desire for a cost effective and robust solution, the focus of the design was on longevity and ease of maintenance.

Manufactured from folded steel, the production methods made efficient use of materials and processing. The laser-cut patterning cohesively links with architectural details on site, while also acting as a skate deterrents and enhanced water shedding. This suite includes backless bench seats and table settings, with the modular design of the tables allowing for a variety of configurations on site.

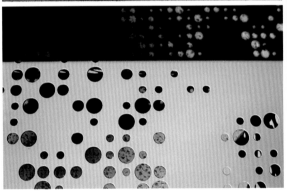

石制户外座椅
Stone Outdoor Bench

设计：Mansilla & Tunon Design: Mansilla & Tunon
完成时间：2010 Year: 2010

FLOR 是一款铸石长椅，具有两种格式，通过比较其相似性与迥异处，进行巧妙安装。该座椅的仿生形态和辐射状形态既适合单独使用，也适合情侣使用。其设计体现出 Mansilla & Tunon 设计机构对平等性和多样性的理解。这款长椅已安装在马德里地区的图书馆和档案大楼的入口庭院中，完善了当地公共集会空间，以开放的姿态欢迎游客共享当地的各种设施。

FLOR is a cast stone bench with two formats, installed using a subtle play of similarities and differences. Its radical biomorphic design permits paired or individual usage to preserve the user's intimacy. The FLOR bench design employs the concepts of equality and diversity, part of the research that has inspired the architectural work of architects Mansilla & Tunon. FLOR was first installed in the entrance courtyard of the Madrid Regional Library and Archives building, the former "El Aguila" brewery. The architects completed the central area of the project with this bench, which has enhanced this meeting point for users of the various institutions that share the complex.

历水湾角阳光海岸座椅
Alexandra Headlands-Sunshine Coast Bench

设计：Surya Graf　　Design：Surya Graf

本案的目的是打造雕塑版的座椅，使其成为历水湾角木板人行道上的一道亮丽风景，与海滩的景色连成一片。因此项目从海滩获取灵感，设计充分地反映出周边的环境和海滩上的生活方式。从整体外观到木板的形状以及弧形的钢架结构，所有构件都反映出海洋的元素。此外，产品的类型也多种多样，有单人座椅，还有公共长椅。所采用的材料包括硬木板和金属零件，并且在制作的过程中运用了计算机辅助设计技术，减少了生产时间，控制了成本。本系列产品具有经久耐用，容易维修的优良品质。

While having an obvious function, the brief for this project was to design a sculptural seating element along the Alexandra Headland boardwalk that would be a focal point while also harmoniously fitting into the foreshore landscape. The design took direct inspiration from its coastal location, reflecting many aspects of this environment and lifestyle. The overall form, the shaping details in the timber and the curved metal framework are all references to aspects of the ocean. The form also allows for individual and communal use of the seat. Relying on CAD technology to produce a quality product in a quick and cost effective manner, the design uses hardwood timber and metal components, which are durable and easy to maintain.

Bench

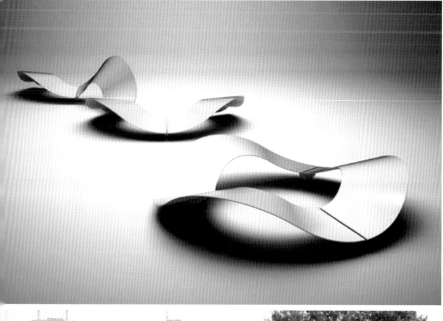
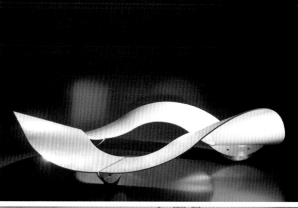

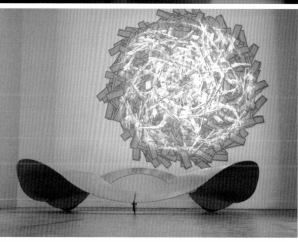

 Ondine 座椅
Ondine Bench

设计：Michael Bihain and Cedric Callewaert
项目地点：比利时

Design: Michael Bihain and Cedric Callewaert
Location：Belgium

迈向无街具的伟大一步。
Ondine 简洁耐用的结构下潜藏的复杂性令人着迷。作为 Parck 设计大奖的得主，该系列座椅的设计理念是：既是雕塑又是座椅。Ondine 由三个模块组成，可根据需要随时调整摆放位置。产品的形状既有长方形，也有圆形。有机的曲线和模块性使得产品具有很强的适应性，无论是城市、文化区域或是乡村地区均可使用。
材料：8 毫米厚不锈钢板
尺寸：外径：254 厘米 内径：125 厘米 最高：57 厘米 最低：17 厘米
颜色：白、浅灰、草绿、天蓝、珊瑚红、深橙、锌黄、金银拼色

"A step towards the NO FURNITURE"
This design tends to fascinate by its underlying complexity translated in a simple and strong form. Laureate of the Parck Design competition, Ondine is the fruit of a close collaboration between Design Michael Bihain and Belgian architect Cedric Callewaert. Ondine's main concept is its capacity to subject its viewe, or user to multiple interpretations——it is both sculpture and bench. Divided into 3 identical modules, Ondine can be repositioned continuously. It offers a longitudinal seating element as well as a circular one. Its organic curves and its modularity offer a wide range of seating possibilities permitting it to continually adapt to various settings, urban, cultural and rural. Material: chromatised powder coated & laquered steel sheets, 8mm / Dimensions: outside diameter: 245cm, inside diameter: 125cm, high seat: 57cm, low seat: 17cm / Color options: white, light gray, anthracite gray, grass green, sky blue, coral red, deep orange, zinc yellow, specially requested silver & gold.

Longo 石基底长椅
Longo Stone Bases Bench

设计：Manuel Ruisanchez Arquitectes   Design: Manuel Ruisanchez Arquitectes

Longo 系列的组成部分包括 Longo Banca 和 Longo Cubo 两个石基、两个木质座椅和一个金属扶手。该座椅分有靠背设计和无靠背设计两种，将所有的部件组装后就形成了一个简单的模块化石椅。由于设计的定位不同，不同的座椅呈现出的外观也不相同。Longo 座椅的基本形状只有一种，颜色有黑色和米白两种可供选择。Longo 座椅直接摆放在地面即可，无需其他固定装置。该系列还有另外两种设计，即 Longo Papelera 和 Longo Cenicero。所有座椅的规格均为 60cm x 100cm x 45cm，另外还有用 AISI 316 不锈钢材设计的垃圾桶和烟灰缸。

The Longo series consists of two combinatory cast stone bases, Longo Banca and Longo Cubo, in conjunction with two wooden seat models and a metal frame with and without backrest. The conjugation of all these components permits alignments of simple modular cast stone benches. At the same time, the seats permit different aesthetic combinations thanks to the possibility of different orientations. With a simple geometry and two colour options, grey and beige, they sit directly on the ground without requiring anchorage. The collection can be complemented with Longo Papelera and Longo Cenicero, both modules measuring 60 x 100 x 45 cm, with a litter bin or ashtray accessory in AISI316 stainless steel.

Bench

柔软座椅
Soft Bench

| 设计：Ucile Soufflet | Design: Ucile Soufflet |
| 项目地点：法国 | Location: France |

座椅令人熟悉的线条在椅面的中点突然波动并向下弯曲，形成一个设计灵活、令人放松的座椅。
座椅的外形仿佛邀请过往的行人坐下、享受美好时光，观察周边的人物和景色。Soft Bench 是理想的街道设计，不仅提供了放松身心的私密空间，同时又与空间与周边景色相映成趣。它是城市里新形态的生活方式。
材料：3毫米厚钢板、表层覆盖锌和环氧粉末涂料
尺寸：长275厘米，宽49厘米，高38厘米
制作方法：激光切割、折叠、焊接

Soft Bench is a bench with a familiar line which suddenly undulates and bends to propose a flexible and relaxed seating on half of its length.
The design asks its user to slow down, take time, and invest in both object and place. It is a new way to appropriate urban furniture and private spaces with the aim of relaxing and contemplating space and landscape. Soft Bench proposes new typologies for city living.
Material: Steel 3mm, treated with primary zinc and epoxy powder coating / Dimensions: L 275cm, W49cm, H 38cm / Process: cut laser, folding, weld

东京座椅
Tokyo Bench

设计：Gehry Partners
项目地点：日本，东京

Design: Gehry Partners
Location: Tokyo, Japan

Frank Gehry 建筑设计公司最近推出了一款蛇形的座椅作为 Design Tide 在东京周末的参展作品。这款淡棕色的长椅目前正摆放在 World Co Aoyama 大厦的大厅里。

The architecture office of Frank Gehry, presented a snaking bench as part of Design Tide in Tokyo over the weekend. Called Tokyo Bench, the maple bench was installed in the lobby of World Co Aoyama Building.

Bench

公共座椅
Public Bench

设计：Raya Hindi
项目地点：澳大利亚

Design: Raya Hindi
Location: Australia

这款以模块化技术打造的户外座椅可以根据场地和使用人数作出不同的安排。

This modular outdoor seating concept allows custom arrangement depending on location and accomodates both group sitting (inside the curve) and individual, more private sitting (outside the curve).

Sucker Punce 座椅
Sucker Punce Bench

设计：Ben Pell
地点：耶鲁大学建筑学院

Design: Ben Pell
Location: Yale School of Architecture

Sucker Punce 的灵感源自耶鲁大学艺术大楼的灯芯绒质感水泥墙。每一块厚度为15厘米的胶合板都准确地固定到墙面纹理之间的缝隙上，形成可供两人并排坐下的空间。座椅在与墙体结合的部分根据需要在水平和垂直方向变得更厚，将粗矿的墙面与光滑的 NURBS 几何表面牢牢固定在一起。

The bench was inspired by the unique corduroy-textured concrete walls of Paul Rudolph——Yale Art and Architecture Building. Each 1/2 plywood profile fits precisely into the spacing of the walls texture (a nod to the original wooden concrete formwork) and articulates a sitting surface for two people. The surface emerges as a thickening of the walls sectional relationship between vertically and horizontality, and juxtaposes the roughness of the wall with the smoothness of NURBS surface geometry.

Tejo Remy 和 Rene Veenhuizen 个人作品展——座椅
Tejo Remy and Rene Veenhuizen Solo Exhibition —— Bench

设计：Tejo Remy and Rene Veenhuizen
项目地点：美国，华盛顿

Design: Tejo Remy and Rene Veenhuizen
Location: Washington D. C., USA

Industry Gallery 在 2010 年 3 月 20 日为 Tejo Remy 和 Rene Veenhuizen 在纽约举办首个个人会展。这个名为 "Hands On" 的活动展出了一系列混凝土浇灌的辅助设施，其体现的设计材料将包括混凝土、竹子、网球和破旧的木垫。

"Hands On" 展出的作品里有一个野餐桌搭配两条凳子，这个组合的灵感来自于关爱儿童健康协会的叶子形状家具。其他展出的作品里还包括用网球做成的管状长椅、为一家中学设计的用狭板做成不规则几何图案形状的 Reef Bench 长椅、两条用回收的木板做成的地毯状的 Accident Carpets、用竹条做成的椅子和一系列用混凝土浇灌的辅助设施。

Industry Gallery will open "Hands On" March 20, 2010, the first solo U.S. exhibition for renowned and innovative Dutch designer Tejo Remy, a founding designer at the Droog Design collective, and Rene Veenhuizen, his design partner of the past decade. "Hands On" will feature approximately a dozen works created from concrete, bamboo, tennis balls, and old woolen blankets.

"Hands On" includes a picnic table and two benches, inspired by the "Leaf Furniture" originally created for a children's health care institution; a large tubular bench made from tennis balls based on a series made for the Museum Boijmans Van Beuningen in Rotterdam; and a geometrically irregular slat wood "Reef Bench" derived from a series created for a secondary school. The exhibition will also feature two "Accidental Carpets" made from recycled blankets, chairs constructed from wide bamboo slats, and a prototype from a new series of poured concrete furniture.

座椅

S形城市座椅
S Urban Bench

设计：Veronica Martinez
项目地点：西班牙，马德里

Design: Veronica Martinez
Location: Madrid, Spain

受字母S的启发，这个城市座椅充分展示了豪美思材质强大的可塑性。该座椅在独自使用或与其他材质结合使用时，不仅可以容纳多人乘坐，而且还保证了外形的美观。设计从对周围的观察中获取灵感，诸如星系的运动、DNA分子链、生长的植物、花朵、流水、飓风、云朵的汇聚，并将其运用到设计当中。
S的造型抓住了两根精致线条运动的本质。有了豪美思的可塑性和坚固性，只用一个模型就可以塑造出S的形状，而且成本超乎想象的低。
S长椅可用于户外，如广场、公园、街道；也可用于室内，如机场、酒店以及会展等。

Inspired by the letter "S" this public bench is a perfect demonstration of the malleability of HI-MACS. On its own or in conjunction with other materials, it can seat a number of people while remaining aesthetically pleasing. The design is based on the shape of her project on spiral motion she observes everywhere, from the infinite elements such as galaxies and chains of DNA to the growth of plants, flowers, how water flows down the river, the movement of hurricanes and the formation of clouds.
The design of "S" is to capture the essence of that movement with two fine lines. Thanks to the malleability and durability of HI-MACS it is possible and "S" can be made with a single mold, whose presentation is perfect with a minimum production cost. "S" can be placed outdoors, such as squares, parks, streets, and indoors, such as airports, hotels and fairs.

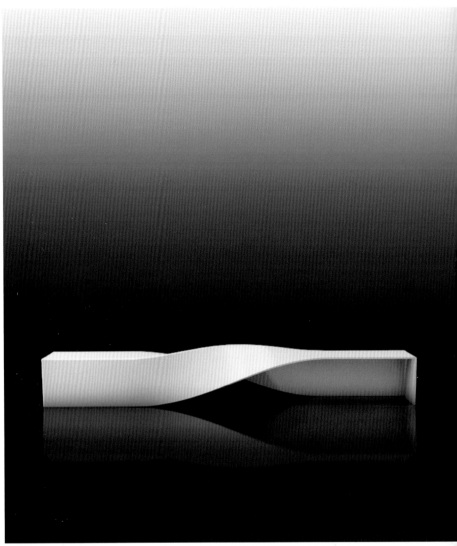

Bench

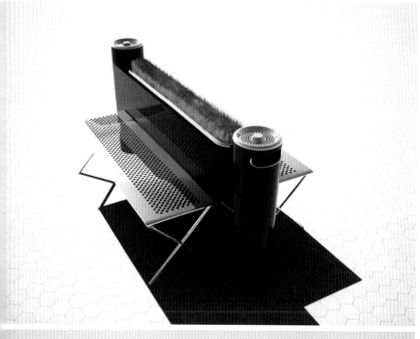

吸烟者座椅
Smoker Bench

设计： Veronica Martinez
项目地点： 西班牙，马德里

Design: Veronica Martinez
Location: Madrid, Spain

把本案命名为"吸烟者座椅"是因为它为吸烟人群提供了一个自由的吸烟和社交场所。两个座位区连接两端的支撑柱，支撑柱顶部是放置烟灰和烟蒂的垃圾箱。支撑柱的高度是座位区的两倍，中间的连接部分种满了青翠的绿草，形成强烈的视觉效果。整个设计以一种温和的方式向周边的人群传递了一种清洁、绿色的信息。它的主要目的是为吸烟人群和不吸烟人群同时提供一个良好的公共环境。

The new project named as Smoker Bench is as the name signifies a place where the smokers can smoke and socialize in peace without a hassle. The bench consists of a two side sitting area with supporting pillars, which double up as containers to flip the ash or disposing the cigarette stubs. The design uses the connecting portion between the two pillars as grass can be grown adding to the visual feel. Also it helps in promoting the clean and green message to all those who spend time there, although in a silent manner. The main aim of this place is to provide a unique public place for smoker and protect the non-smoker at the same time.

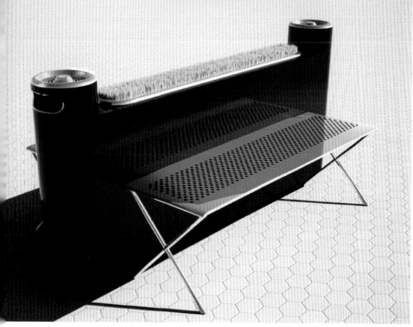

座椅

自行车式街道设施
Bike-Inspired Street Furniture

项目档案	Project Facts
设计：Jihyun David	Design：Jihyun David
项目地点：荷兰，阿姆斯特丹	Location：Amsterdam, Dutch
完成时间：2010	Year：2010

本案的主要目的是充实和活跃整个街道。本案是一个临时的街道设施，其设计灵感来源于荷兰的自行车文化。座位采用荷兰典型的自行车车座，并且固定在稳固的金属护柱上。护柱的底端有一个搁脚板，不仅方便而且很舒服。

The project is meant to activate the street as a place to enjoy. The project is a temporary street installation inspired by the Dutch bicycle culture. The seats are made from classic Dutch bike saddles and constructed on the bollards with a solid piece of metal. At the bottom a footrest is attached to the bollard to make the stool pretty comfortable.

Bench

座椅

散步广场
Promenade Square

项目档案

设计：Tonkin Liu
项目地点：英国，曼彻斯特
完成时间：2011

Project Facts

Design：Tonkin Liu
Location：Manchester, UK
Year：2011

本案将城市中曾被忽略的一角转变成一个动态的城市公共空间。在这里，人们可以得到无比轻松自在的享受。他们可以坐着、漫步、吃东西、聊天和娱乐。18棵英国梧桐树加入了原来的21棵树木中，绿化更加到位。石头路面上有序地排列着暗灯。在树木的周围也设计了圆形的护栏，不仅可以保护树木，也为人们提供了休息的地方。石头地面砖大小一致，方便维修或更换。照明是由太阳跟踪计时器开关控制的，高高的射灯可以发出6至8个方位的光线。

The project has transformed a neglected part of the city into a dynamic routing and public space where people can sit, linger, eat, talk, play and meet in a relaxed manner. 18 London plane trees were added to the existing site of 21 mature trees and a stone pavement light measuring 2015 square meters was laid down. Shifting the traffic circle around the trees in clusters and clearings of the promenade are encouraging the various rhythms of its users and create rooms in urban areas away from trees. Stone panels are pre-cut and designed into the slot of the same puzzle pieces around furniture for easy maintenance or replacement. Lighting operated by a sun tracking timer switch is provided by columns of tall lamp with 6-8 multi-directional beams.

Bench

石头状的座椅
Stone Bench

设计：Peter Donders
项目地点：比利时
完成时间：2010

Design: Peter Donders
Location: Belgium
Year: 2010

这一系列座椅是由比利时的设计师创作的。首先将一根碳纤维一端在模型上固定，然后沿着模型表面不停地缠绕，完成后将中间的模型抽掉，形成眼前这个既现代又有机的结构。完成后的座椅透气性良好，并且非常坚固，《花园杂志》的 Rob Cassy 称其为"三维的书法"。目前，碳纤维这种轻质而且坚固的终极材料被用作F1方程式赛车和宇宙飞船底盘的生产材料。本系列座椅共有10件，既可用于公共场所，也可放置在私人空间内。
Stone 系列座椅的制作方法和材料如 Bench 系列如出一辙，外形就像它的名字，看起来就像一块静止的石头。该系列也只生产了10件。
材料：碳纤维
尺寸：长：300 厘米 高：45 厘米 宽：60 厘米
重量：6 千克

The Belgian designer created this contemporary and organic piece by twisting a single string of carbon fibre around a form that was then removed. The resulting structure is airy yet incredibly strong and has been aptly described as "Calligraphy in 3D" by Rob Cassy of the Garden Design Journal. Currently the ultimate material available in terms of weight to strength ratio, carbon fiber is used to produce Formula One race cars, the highest quality sporting equipment, as well as the chassis of space crafts. This exceptional limited edition series will be limited to 10 pieces and are suitable for both public and private spaces.
Inspired by its namesake, Stone is created with the same eye for detail, depth and movement as Bench. As a unique sculpture and seat, Stone is produced in a limited edition of 10 pieces.
Material: carbon fiber / Dimensions: L 300cm, H 45cm, W 60cm / Weight: 6 kilos

Bench

BD Love 座椅
BD Love Bench

设计：Ross Lovegrove	Design: Ross Lovegrove
项目地点：比利时	Location: Belgium
完成时间：2010	Year: 2010

这款 BD Love 座椅，可用于家居、街道、花园、机场、酒店门厅、酒吧或者鞋店中。Ross Lovegrove 设计的公共座椅具有生机勃勃的雕塑感，外形饱含了设计情感。这款长椅具有宽大的尺寸，可提供 10 个座位。

座椅的材料采用的是聚乙烯，并使用滚塑工艺生产技术制作，座椅的色彩可在荧光红色、米色、白色、蓝色、绿色、砂岩色和磨石色（暗灰色花岗岩）等范围内选择，也可根据客户的要求定制特殊颜色。（尺寸：长 2652X 宽 1293X 厚 410X 高 941/ 高 1761mm）

BD Love benches, the urban transit furniture concept for transit spaces, are designs that are equally at home, on the street, in the garden, in an airport, a hotel foyer, a bar or indeed a shoe shop. Ross Lovegrove has created a public seating that has all the character of exuberant, vibrant sculpture, with forms designed to engage the emotions. Its generous dimensions allow it to seat up to ten people at once and children to have a whale of a time.

The bench is made from rotation molded polyethylene and is available in fluorescent red (not for outside use), beige, green, black, sandstone and millstone (dark grey) or client's own colour specification with minimum of 12 units. (Dimensions: L2652xD1293xSH410xH941/H1761mm)

座椅

 座椅和广告板两用系列
Bench & Billboard

设计：David Szabo	Design: David Szabo
项目地点：澳大利亚新威尔士	Location: NSW Australia
完成时间：2011	Year: 2011

广告板的出现丑化了城市的景观，又挡住了风和光照，而且由于材料几乎都是木材和纤维，对环境造成的影响也十分恶劣。但由于广告板的应用在经济发展过程中是符合市场发展需求的，所以即使不喜欢，公众还是得忍受。而今，David Szabo 提出的座椅和广告板两用的创新概念为公共场所广告运用提供了一个新的方向。新型座椅的屏幕可以用来展示广告、图片、录像和其他宣传，就像普通的广告板一样。而和普通的报摊平面广告不同，新型座椅将邀请行人与之互动。例如，当行人经过或是站在座椅前方时，座椅上的屏幕开始缓慢地展示宣传信息，并发出亮光，以此吸引行人的注意，同时展示产品和服务信息。当行人停在距离座椅 70 至 200 厘米的距离并超过 5-10 分钟的时候，它便会自动进入座椅模式，关掉显示屏和其它灯光显示并向前方以任何角度弯曲，形成一个座椅。这看起来似乎比普通的座椅都要麻烦，而且广告板似乎也成倍增加了，但是这种设计却充分考虑到了空间因素，不用的时候可收起来，完全不占空间。因此，这类座椅极适合用在人流量集中的地方，如公交车站、加油站、火车站、会展中心等场地。

Not only do they deface good urban landscape, the billboards also block the movement of wind and sunlight, and being crafted out of paper or some other fabric derived out of wood, and also super environmentally-unfriendly. But the billboards have become an important part of the economic process these days and have to be endured by the general public even if they don't like them. Luckily, a few innovative designers are doing their best to make this form of advertising as functional and un-invasive for the public as possible. The kind of seating concept by Design David Szabo is one such initiative to add another dimension of function to advertisements in public spaces. The bench is a piece of public furniture that is a billboard screen and bench at the same time. Its screen can be used to display advertisements, graphics, videos and so on, and can be let by companies and brands just like normal billboards. The Bench, however, would differ from your regular kiosk-style billboard in the way the public interacts with it. For example, if one walks or stands before the Bench, the info on the screen starts to move softly and its electronics spring into action attracting the person's attention and giving them the information about the product and service that they are advertising. After the person has stood in front of the screen at a distance of 70 cm to 200 cm for 5-10 minutes, the Bench then automatically goes into furniture mode and switches off its screen and other electronic displays and bends in front of the person, in either direction, and transforms into a seat. This may seem a little too much hassle than a regular public bench that also doubles up as billboards at times, but it makes a lot more sense for places with less space on the sidewalk since the bendch folds up when not in use allowing the space around it to be used for regular movement of people. The installation is also very useful for public spaces that are prone to crowding like bus stops, petrol stations, railroad stations, exhibitions, etc., which need seating to be offered as a regular feature but also on occasions need all the available space that is often taken up by fixed seating.

Bench

座椅

Section2 系列座椅
Section 2 of Street Furniture

设计：James Corner
项目地点：美国纽约高线公园
完成时间：2010

Design: James Corner
Location: New York City's High Line Park
Year: 2010

Section2 系列座椅独具个性，周边是茂密灌木、野花、高架立交桥和绿草如茵的草坪，是逃离水泥建筑丛生的城市喧嚣的最佳场所。

Section 2 definitely has its own personality. With a thicket of trees, a wildflower field, an elevated flyover, and a lush grassy lawn, Section 2 is truly a unique escape from the hustle and bustle of the concrete jungle.

Bench

乡村公用座椅
Rural Township Bench

设计：Surya Graf 　　　　Design: Surya Graf

本案是专为 Maleny 小镇设计的街道辅助设施。该座椅使用预先定形的木板打造椅身，然后用钢折板将各个部件连接在一起，既保留了传统的工艺，又展现了当地的文化。该系列产品的类型包括靠背式座椅、无靠背座椅、台式座椅和桌子等。

This range of custom furniture items was designed specifically for the Maleny township site. While the area is rapidly expanding, it continues to support a culturally alternative community. In keeping with the themes of traditional craft arts and with forms inspired by the rural context of the location, the furniture features shaped timber that has been combined with folded steel components. This suite includes a variety of backrest bench seats, backless benches, platform benches, and table settings.

Bench

Dutton 公园座椅
Dutton Park Bench

设计：Surya Graf   Design: Surya Graf

在Wilson Architects的协助下，本案成了连接昆士兰大学与Dutton公园的纽带。宽大的座椅以不锈钢部件和硬木板打造，既适用于公园这类自然环境，也适用于大桥和公交车站这类现代建筑丛林。本案还应用了许多昆士兰大学现有的设施，如垃圾桶、喷泉和广告板。

Designed in conjunction with Wilson Architects, the furniture for the Green Bridge development was intended to be a linking element between the University of Queensland site and the Dutton Park council parklands. Constructed from stainless steel sections and hardwood timber slats, the large custom benches were designed to sit well within both the natural setting of the parklands and within the modern architectural lines of the bridge and bus terminus. This project also utilized many of our existing products previously specified throughout the UQ campus, including bin enclosures, drinking fountains and bollards.

座椅

意大利废弃铁轨改造项目
Retired Italian Railroad Transformed

设计：Bridgette Meinhold
项目地点：意大利沿海岸 Celle Ligure
完成时间：2011

Design: Bridgette Meinhold
Location: Celle Ligure on the coast of Italy
Year: 2011

今年夏季，位于意大利沿海城市 Albisola Superiore 和 Celle Ligure 之间的一段废弃的铁轨被打造成休闲的散步道。曾经毫无用处的海岸废墟摇身一变成为当地用途最大、最受欢迎的旅游热点，既具休闲氛围又可以欣赏海洋美景。本案的目标是打造对环境敏感的空间，同时保证将材料对环境的影响降到最低。

为了达到散步道的成功转型，设计必须进行一系列的改进，从而使项目在休闲空间和旅游景点这两个方面占据优势。设计的重心是为散步道赋予连续性，提高海岸线的使用率；使用更加环保的材料；提高公园的人流量，为旅游产业创造新的出路。项目从头到尾都贯彻了保护脆弱的生态地理环境和美丽的自然美景的环境理念。而铁轨的改造也将重心放在了如何提高隧道和礁坡的安全性上面。散步道沿岸的自行车道和人行道对外公开，座椅和休闲区为公众提供了欣赏美景的良好环境。

A new recreational promenade was created out of a section of retired railway in between the towns of Albisola Superiore and Celle Ligure on the coast of Italy. What was once an unused cut in the coast is now a useful and prominent tourist attraction for the area, which provides space for recreation and views of the ocean. The project is designed to be as environmentally sensitive to the space as possible while making use of low impact materials.

Converted to a walking path, the promenade also includes a creation of vantage points that would encourage its use as a recreational space as well as a tourist attraction. The main goals of the project are to provide continuity to the pedestrian paths, increase the usability of the coast line, make use of materials with low environmental impact, increase use of public parks, and create new opportunities for tourism. The entire project is carried out in an environmentally aware manner to protect the sensitive site and beautiful vistas. The renovation of the railway also focusses on improving the safety of the tunnels as well as the reef slope. Bike and pedestrian paths are open and accessible to the public and benches and rest areas are provided for those who want to enjoy the view.

座椅

Piano 系列
Piano Street Furniture

设计：ADDI Company　　Design: ADDI Company
项目地点：美国　　　　Location: USA

Piano 系列是自行车停车架和公共座椅结合的产物。90 度弯曲的造型是整个环境的焦点。该系列的座椅有三种变形，既节省空间，外观也十分漂亮。这个产品一半是座椅一半是单车架，随着对生态运输的要求日益加强，该系列产品为解决这一城市问题提供了很好的解决方案。

Piano is combined bikestand and park bench. The 90 degree angels creates an interesting and effectfull contrast in the landscape. The bench will come in three different variations. This piece created by ADDI saves space while looking beautiful: it is part bench and part bike rack. With the rising demand for ecological means of transport, this bench is perfect for any city.

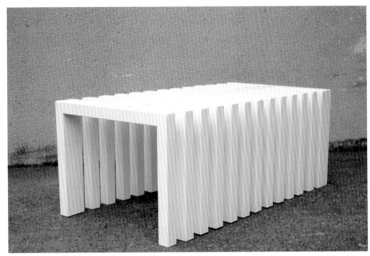

Bench

 Toast 4 沙发
Toast 4

设计：ADDI Company	Design: ADDI Company
项目地点：美国	Location: USA
完成时间：2010	Year: 2010

Toast 是适用于公共场所的一个非常好玩的沙发，特殊的结构使它能够适应各种大的小的、高的矮的的墙面。Toast 还具有雕塑般的外观，从任何角度看过去都十分有趣。

Toast is a playful sofa for public spaces. It's been designed to fit short, tall, big and small. The sculptural shape makes Toast interesting from all angles.

座椅

VIAS 空间
VIAS Space

项目档案

设计：estudioSIC, Esa-Acosta, Mauro Gil-Fournier
项目地点：西班牙
面积：1 450 平方米

Project Facts

Design：estudioSIC, Esa-Acosta, Mauro Gil-Fournier
Location：Leon, Spain
Site Area：1,450 m²

重建旧的铁路大厦——这个项目意味着人们愿意投入更多到城市的结构和建筑当中，为年轻人提供新的娱乐设施，在市中心再造一个新的公共空间。可移动性、建筑、社会和城市可持续性发展的关系为这个项目奠定了基础，降低了成本，最大化地利用所有的资源。原来半废弃的面貌经过合适的改变后，增加了空间的可达性以及活动的丰富性。
安全、灯光以及声音等都是公共活动成功展现的重要因素。这个空间并不是呆板的，它灵活性很强，而且可以容纳很多人。特别是有一些隐藏的功能让这个空间不仅很好地传递了文化元素，而且还增强了这个空间的网络结构。

Bench

VIAS case means a strong investment for people to get involved in the urban structure and constitution. The project restores an ancient rail building for art production, provides a new brand equipment for the youth and regenerates a new area in the city centre through the recovery of FEVE tracks as public space. The relations between mobility, architecture and social and urban sustainability set the basis for the development of this project: minimizing resources and taking advantage of the opportunity to work together between management, architecture and city. The semi-derelict appearance that the old building, also known as "cocheras" of FEVE presents makes necessary a moderate intervention, in order to allow the public access as well as the correct development of any activity.

This intervention must only be considered as the restoration and recovery of what already exists, just adding logical elements in terms of storage, health and safety, lighting and the sound system that the schedule of activities will require. Flexibility, accumulation and undetermined function make espacio VIAS project a very close space to develop lines of action that should help design cultural strategies, settled on the aforementioned infrastructure and to be expressed into citizen activities of cultural promotion and the creation of networks according to young public.

座椅

拉斯内格拉斯滨水区座椅
Las Negras Waterfront Bench

项目档案

设计：Jesus Torres Garcia 建筑事务所
项目地点：西班牙

Project Facts

Landscape Architecture：Jesus Torres Garcia
Location：Cabo de Gata Natural Park, Almeria, Spain

拉斯内格拉斯滨水区的景观设计包含了很多公共元素，设计的选材旨在加强城市元素之间的互动，木头的结构、围边座椅上的覆盖，设计还融入了大海的声音、自然环境的形状、地形的排布和附近的植被，这些都是为了鼓励和巩固公共元素的呈现和实施。

The example of Las Negras elicits a considerate approach to the public element; the choice of materials has been decisive in consolidating the work within its urban interaction; the wood of the structure and the coverings of perimeter benches encourage a pleasant treatment favouring its consolidation. There is a symmetrical psychology in the human treatment that is here applied to the use and the form of the material as a means of the object's expression. This reflection refers to the natural element, the sound of the sea, the material and shape of natural elements, the vegetation, the geological configurations, as well as the settings of interest.

Bench

座椅

  意大利面条长椅
Huge Spahetti Bench

设计：Pablo Reinoso
项目地点：意大利
完成时间：2011

Design: Pablo Reinoso
Location: Italy
Year: 2011

阿根廷裔法国设计师 Pablo Reinoso 设计了一款"Spahetti Bench"（意大利面条长椅），是一款类似雕塑的木制长椅。Sudeley 长椅是设计的另外一款的特色长椅，既是户外雕塑，又充当了长椅。这种长椅总共设计了 8 张，每张椅子都采用钢铁制作，长为 9 米。

This is a very sculptural wooden bench entitled "Spahetti Bench", a small focus of another creation by Design Pablo Reinoso from Argentina and now based in Paris, France. Halfway between the bench and outdoor sculpture, the "Sudeley Bench" is an impressive creation in steel, 9-meter long and only 8 copies.

Bench

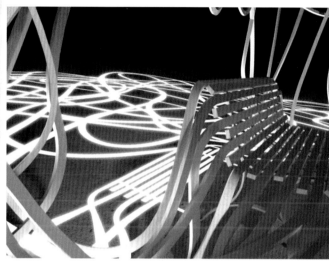

座椅

Mount Fuji Architects Studio 设计的长椅
Mount Fuji Architects Studio Bench

设计：Mount Fuji Architects Studio
项目地点：日本十和田艺术中心
完成时间：2010

Design: Mount Fuji Architects Studio
Location: Towade Art Centre in Japan
Year: 2010

这些镜面不锈钢长椅是由日本设计公司 Mount Fuji Architects Studio 设计的，摆放在十和田艺术中心的樱花树下。每张长椅由两个或两个以上的不锈钢原件折叠构成，并在椅面上进行了镜面处理，光可鉴人。十和田四季景色秀美，冬天雪花纷飞，春天樱花盛开，夏天阳光透过树叶在地面上留下斑驳树影，秋天落叶纷飞。镜面的长椅倒映着这些美丽的景色，形成"景在镜中"的唯美效果。

These mirrored benches by Japanese firm Mount Fuji Architects Studio sit beneath cherry trees at the Towada Art Centre in Japan. Each bench comprises two or more folded stainless steel elements joined together to support each other. Towada city has snowflakes floating around in winter, cherry blossoms floating around in spring, sunlight sifting down through the trees in summer, leaves floating around autumn, and artworks spreading over the city. There is always something floating around through the city. Its polished surface like a mirror reflects many kinds of things in Towada city.

 ### 促进交际的另类会面碗
Offbeat Meeting Bowls Promote Community

设计：public art studio	Design: public art studio
项目地点：纽约	Location: New York City
完成时间：2010	Year: 2010

今年夏天，与其在拥挤的时代广场穿行，何不坐在会面碗中小憩片刻？时代广场为人们营造了一个另类的社交空间，可供朋友、甚至陌生人坐在碗中，享受休闲时光。会面碗是一款酷炫而又时尚的家具设施，放置在时代广场中心地带，每个碗可容纳8个成年人，均采用回收数控切割硬纸薄板制作，营造了一个亲密而又开放的社交空间。

This summer, instead of pushing through the summer crowds in Times Square, why not take a break and hang out in a Meeting Bowl? The Times Square Alliance has teamed up with Spanish collaborative to create a social space for friends and even strangers to gather and chat, promoting dialogue and good old fashioned friendliness. As cool, temporary urban furniture installations, these quirky bowls serve up some much needed calm in the middle of the madness of Times Square. The three giant bowls rest on a base that gently rocks to provide a mild floating sensation and each bowl can seat up to eight adults. Made of thin slabs of recycled CNC cut fiber board, the bowls are at the same time an intimate, yet remain open to the views of the city.

Vekso 座椅
Vekso Bench

设计：part of NRGI group	Design：part of NRGI group
项目地点：德国	Location: Germany
完成时间：2010	Year: 2010

Vekso 长椅、座椅和桌子都采用最好的材料精心加工而成。这些材料都是精心选择的，可经受风吹雨打、日晒雨淋，兼具功能性和观赏性，随着时光流逝，与周围的环境融为一体。底座：切割花岗岩，抛光或喷射燃烧处理；钢铁：热浸镀锌钢。也可进行粉末涂层处理；扶手：座椅也可配备热浸镀锌钢制扶手。

All Vekso benches, seats and tables are made with care and craftsmanship from the finest materials. All Vekso materials have been specially selected to withstand wind and weather and for people who value form and function. Vekso uses high-quality metal and wood capable of withstanding constant use. Some materials require maintenance. Others age with grace, gradually developing a natural patina that allows the product to blend with its surroundings. Base: Cut granite can be delivered as polished or jetburned. Steel: Hot-dip galvanized steel and powder coated steel. Arm rests: Outfit seat is also available with arm rests in hot-dip galvanized steel. when ordered together with seat.

Bench

 Rodeo 座椅
Rodeo Bench

设计：part of NRGI group	Design: part of NRGI group
项目地点：德国	Location: Germany
完成时间：2010	Year: 2010

Rodeo 长椅兼具多功能性与美丽外形。这款长椅不分正、背面，两面都可以坐，可以单张椅子摆放，也可以多张椅子组合摆放，用于适应不同的需求。主要材料采用的是硬木和不锈钢，便于修护。这款椅子既可用于公共场合，也可摆放在私人空间，是一款非常实用的长椅。

The Rodeo Bench, symmetry of form and versatility, can be positioned to make the best of two views, with users facing forwards or backwards. Rodeo is freestanding but can be fixed with a minimum of preparation. Fabricated in tough hardwood and low maintenance stainless steel, Rodeo performs equally well in public and private settings.

北湖住宅区座椅
North Lakes Estates Bench

设计：David Shaw & Surya Graf
Design：David Shaw & Surya Graf

这款长椅是为北湖住宅区量身定制的。其外形源自住宅区的建筑形状，独特而又与住宅区的花园、社区空间和谐统一。设计希望这款长椅拥有实用、寿命长、易于维护的特点，因此采用的主要材料为浇铸铝元素、预铸混凝土及硬木木板。

This custom furniture suite was designed specifically for the North Lakes residential estates. With forms inspired by the architecture on site, the main focus of the furniture suite was to create a unique and unified identity for the parkland and community areas of the estates. Motivated by longevity and ease of maintenance, the initial material palette for the suite mainly included cast aluminium elements combined with pre-cast concrete and hardwood timber.

Bench

座椅

Zero Collection 座椅
Zero Collection Bench

设计：Doro Design
项目地点：意大利，托里诺
完成时间：2011

Design: Doro Design
Location: Torino, Italy
Year: 2011

Zero 系列家具代表了 Doro Design 公司的本质：Zero 系列家具是 Doro Design 生活方式的本质。这系列的椅子采用铝制作，铝是一种线状清洁材料，可根据需要塑形，以代表 Doro Design 公司的灵魂。简洁的切割线条形成柔软而富有生命力的家具，反映了精细、奢华与都市风格的统一。Zero 椅子仅由一块铝板制成，通过 3 个切割面、2 次折叠，展现了极简设计。长躺椅和衣帽架的设计延续了简洁美丽的设计线条。

The Zero Collection represents the essence of the firm——Zero Collection is the essence of the Doro Design lifestyle. Designed entirely around the aluminum linear and dean material able to transform into tangible shapes the soul of the studio. Precise cuts and lines that suddenly find themselves soft, simple but full of vital energy, the Zero Collection reflects the fusion between the sophisticated and urban, luxury and metropolitan. With just three cuts and two folds on a sheet of aluminum, the Zero Chair is a beautiful example of simple and minimalist design. The chaise longue and coat rack follow the same cut-out simplicity with beautiful lines.

Bench

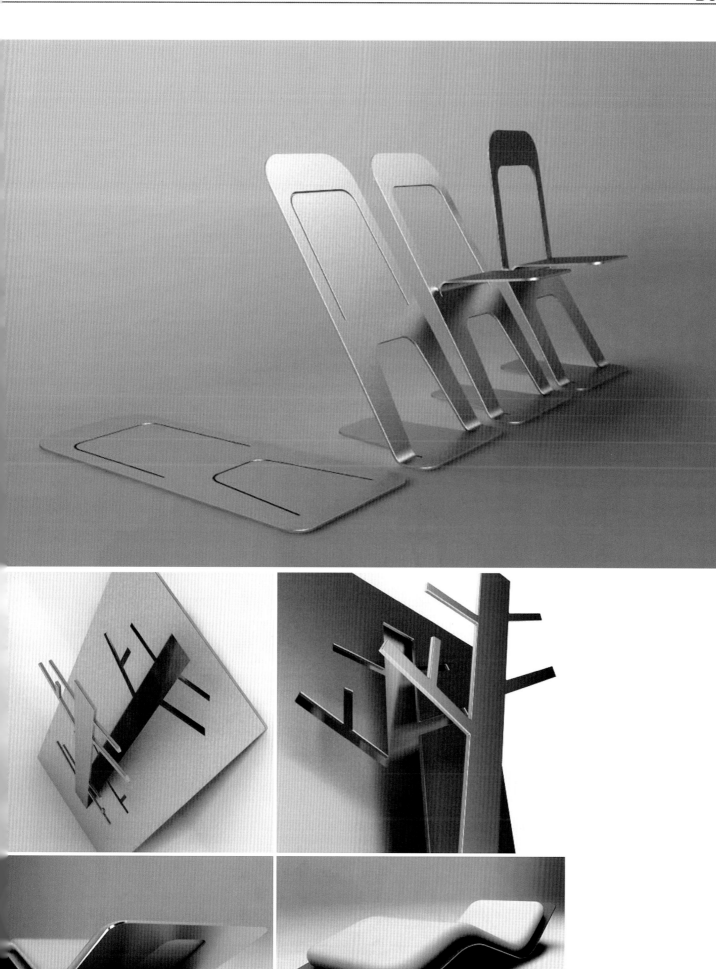

座椅

信纸座椅
Letter Bench

设计：Nicolausson Martin and Tom Eriksson

Design: Nicolausson Martin and Tom Eriksson

这款长椅采用硬木制作，配上不锈钢框架，整张椅子采用一整块木块制成，并未包含任何接缝，这样的设计给人一种错觉：从椅子前面望去，这张椅子仿似在空气的压力下自己折叠了。这款椅子位于Bristol的一座医院里，椅子的设计原型来自医院的一位病人所写的明信片。由于英式英语与美式英语的不同，这封信读起来令人感觉既平凡又奇妙。

The postcard-inspired wooden bench comes with its very own postage stamp in the proper corner, creased into a bend that replicates a folded piece of mail. Constructed of hardwood, Letter Bench is supported on a stainless steel frame, even though the piece creates somewhat of an optical illusion: from the front, it looks like the bench is held up by nothing but air. Installed at a hospital in Bristol, Letter Bench presents a hand-written letter from a patient who previously stayed at the hospital. Due to the differences in British and American English, the actual text reads like something completely mundane or utterly fantastic.

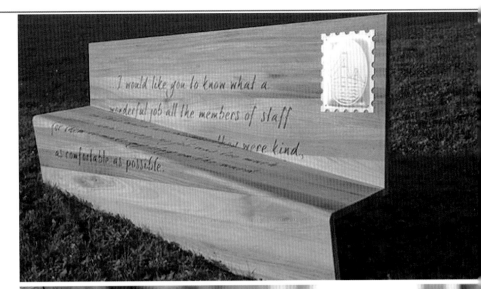

跷跷板式长椅
Seesaw Bench

设计：Nicolausson Martin and Tom Eriksson

Design: Nicolausson Martin and Tom Eriksson

这款长椅由青年设计师Nicolausson Martin 和 Tom Eriksson 设计。它是为了促使陌生人在公共场合畅谈而设计的，需要两人保持默契才能发挥其功能。

Designed by young designers Nicolausson Martin and Tom Eriksson, the concept of "Seesaw Bench" is an attempt to encourage people to discuss them in public places. This bench requires full cooperation to be functional!

Bench

Energy-Moke 座椅
Energy-Moke Bench

设计：Daniel Abendroth	Design: Daniel Abendroth
项目地点：德国	Location: Germany
完成时间：2011	Year: 2011

未来能源将会愈加稀少，愈加昂贵。这个项目的存在就是为了解决这一问题。Energy-Moke 座椅是一款自行车样的椅子，放置在公共场所，可为使用者的移动设备充电。这款绿色加白色的椅子是由豪美思制作的，顾客也可以根据需要自由选择不同的颜色。椅子前盖设置了 USB 连接器，用于插入移动设备，有了这款椅子，外出时，人们再也不用担心移动设备没电。

Energy will be more scarce and expensive in the future; this project delivers solutions to simple but important problems. Energy-Moke is an innovative static bike-like object for public places, which allows users to recharge their mobile devices while cycling. It is made from HI-MACS in white and green colours, but of course it can be customized, thanks to the wide range of HI-MACS colours available, to match the colour identity of companies or cities. The front cover has a USB connector where you plug in your mobile device. Striking and savvy, Energy-Moke puts an end to worrying about running out of battery!

Ensemble 座椅
Ensemble Bench

设计：Roel Vandebeek
Design: Roel Vandebeek

街道家具生产商 Wolters 邀请设计师 Roel Vandebeek 为他们设计了这款座椅。简单的线条勾勒出现代简洁的 Ensemble 座椅，而白色底座仿似阳光照射下，椅子投射在地板上的白色影子。

The Ensemble bench was created by Roel Vandebeek for the urban street furniture manufacturer Wolters. Simple lines were used to create this beauty but the contemporaneity comes from how Roel managed to think a network of stands on which the bench rests. It looks as if white shadows were cast in the process.

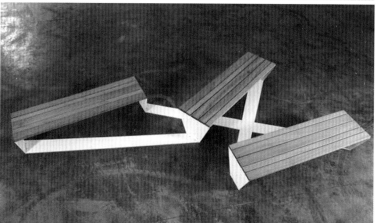

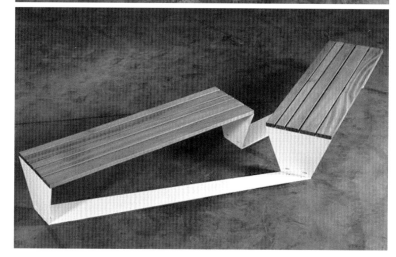

座椅

 悬臂式长椅
Cantilevered Bench

设计：Stokke Austad　　Design: Stokke Austad
项目地点：挪威　　　　　Location: Norway

这款悬臂式长椅可根据需要自由组合成不同的椅子。挪威可持续木材生产商 Kebony 授权设计师 Stokke Austad 设计了这款长椅，用以探索证明应用工程软木制造户外家具的潜力。随着岁月流逝，木材仍将保留耐磨的物理特性。

The bench is part of a seating system that allows several benches to be connected in a variety of arrangements. The project was commissioned by Norwegian sustainable timber producers Kebony to explore and demonstrate the potential of using their engineered softwoods in outdoor furniture. The timber will gradually age during its time exposed to the elements but the designer say it will maintain its hard-wearing physical properties.

Bench

座椅

 Muungano 座椅
Muungano Bench

设计：Peter Thuvander
项目地点：瑞典斯德哥尔摩国家画廊

Design：Peter Thuvander
Location: National Gallery in Stockholm, Sweden

本案是瑞典设计师 Peter Thuvander 专为斯德哥尔摩国家画廊"概念设计"艺术展精心准备的。其中，Muungano 是一个钢板凳，它的意思是"联盟"。

Designed by Swedish designer Peter Thuvander, a reflection on the language and forms for the exhibition "Conceptual Design" held at the National Gallery in Stockholm, the concept "Muungano" is a steel bench that is laser cut. Its name means Union!

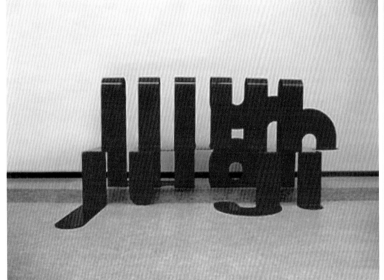

Bench

弧坑座椅
Crater Bench

设计：Ilkka terho	Design：Ilkka terho
项目地点：英国	Location: UK
完成时间：2011	Year: 2011

Crater Bench 是 Ilkka terho 专为 Peter Pepper 设计的座椅。该座椅通常有2-3个坐部，具有轻质、耐用等优良性能，而且室内、室外都适用。弧形的曲面和圆润的边缘让它在众多座椅中脱颖而出，可以调节高度的凳腿让调整设计变成小菜一碟。该类座椅具有防辐射、防潮、防腐蚀、防温度变化等优势，聚乙烯制造的本体有4种颜色可供选择，且100%可再生。

Crater Bench is designed by Ilkka terho for Peter Pepper. Offered with 2 or 3 seats, the Crater Bench is lightweight, yet durable for interior and exterior use. Contoured seating areas and soft edges make this an excellent choice in multiple bench seating and adjustable foot glides make re-arranging your design a breeze. Crater Bench is resistant to UV, moisture, corrosion and temperature changes. It's made of polyethylene, offered in 4 colors, 100% recyclable.

Botanist 系列桌凳
Botanist Table and Bench

设计：Evelyn Lee	Design：Evelyn Lee
项目地点：美国	Location: USA
完成时间：2010	Year: 2010

Botanist 户外系列拥有四种不同的装饰，可以经受住户外环境的恶劣气候条件，室内系列则采用木饰面装饰，让你既能享受大自然的舒适又能将室外景色融入室内。不同的长度和颜色组合能够满足各种要求，无论是在居室还是后院都能找到最完美的搭配。

With four finishes durable enough to weather the outdoors and two natural wood veneers for the interior of the house, Botanist gives you a comfortable way to enjoy the outdoors or bring some of the outdoors in. The varying different lengths and color combinations allow you to mix and match to your heart's content, finding the aluminum was recycled, instead of just recyclable.

座椅

 Hello Stranger 系列座椅
Hello Stranger Bench

设计：Peter Thuvander	Design: Peter Thuvander
项目地点：瑞典斯德哥尔摩	Location: Stockholm, Sweden
完成时间：2010	Year: 2010

Hello Stranger 系列座椅以凹陷的造型鼓励人们积极交流，同时坐席区的分开设计又确保了私密的个人空间。Hello Stranger 座椅可根据场地需求创造出有趣的造型，鼓励各种各样的互动。座椅本身用一个坚固的不锈钢架和一个坐部打造，坐部的材料有多种选择，包括硬木板、金属、甚至是塑料。

Hello Stranger's concave form encourages sociability but leaves open the option to sit apart. Used in multiples, Hello Stranger can be used to create interesting layouts that encourage different forms of interaction. Hello Stranger consists of a tough stainless steel frame together with a seat insert available in a variety of materials, from hardwoods to metals and plastics.

Philly Pods 座椅
Philly Pods Bench

设计：Arman Dhowdhury, Leed Ap and Ms Arch Upenn

Design：Arman Dhowdhury, Leed Ap and Ms Arch Upenn

未来，新式的现代设备、有效的交通和环保的高层建筑将为城市人口带来更多的便利。Philly Pods 就是这样一款设计，不仅可以给费城大街增添一抹亮丽的风景，而且还为当地居民提供了一个休闲的好去处。模块式的设计可以让座椅根据场地现状和用户需求随时改变自身的形状。为了给用户提供更加舒适独特的体验，座椅还增添了弹性设计，不管是坐还是站在凳子上的感觉都很棒。到了夜晚，座椅的边缘被灯光照亮，整个街道的夜景显得愈发迷人。白天还可以在座椅上方安装一把遮阳伞，这样就可以避免太阳暴晒。

When future civilization is expected to be facilitated with ultra-modern appliances, effective transportation solutions and eco-friendly high raised buildings, Philly Pods have been designed to give Philadelphia streets a unique appearance and the inhabitants a great place to seat. This monolithic seat can be arranged in a variety of patterns to meet up different needs and number of the users. To make the seating unique and comfortable, it features springy-ness which gives an extraordinary feel to the users when seating or standing up on it. Moreover, these seats can be lightened during evening through rims of translucent Corian that complements the overall appeal of the streets. Also, a shade can be incorporated over the seat during daytime to provide some protection from the burning sun.

Works of OMC Design Studios

项目地点：意大利　　Location: Italy
完成时间：2011　　　Year: 2011

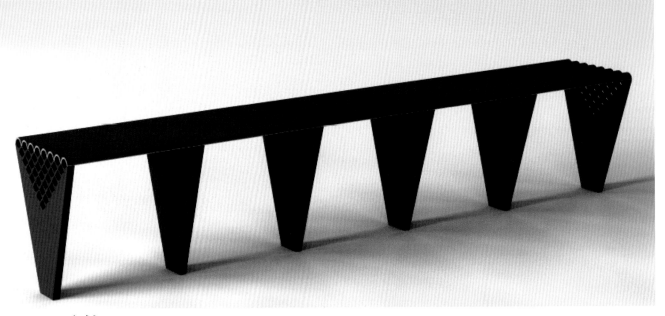

 Dantelli 座椅
Dantelli Bench

Dantelli 座椅的灵感源于网状花边。该座椅既时尚又实用，可同时容纳 5 个人。制作方法：以钢管打造椅面，以激光切割塑造椅身，以焊接的方式连接所有部件，最后安装水泥底座，将整个座椅与地面固定。
材料：金属、混凝土

The Bench: The Dantelli design takes its inspiration from the laceworks, and is both decorative and functional, providing seating for at least five people. How to Produce: This design could be classified as semi-ready-made; the seating is the tubular stainless steel pipes that people use in buildings. And the bottom is laser cut to the shape. They are fixed together by welding, and the bench is fixed to the ground by a concrete root. Materials: Metal, Concrete.

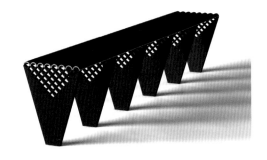

 Hammock 座椅
Hammock Bench

Hammock 浇铸金属座椅以吊床为原型，是现代街道和旅游景点的最佳选择。
制作方法：纯金属浇铸，座椅的原模必须详细、精确，边角圆润。
材料：金属

The Bench: Totally inspired by hammocks, this casted metal bench is no like other, and it is perfect for modern streets and touristic locations. How to Produce: The object is metal casted with a high-detailed mold with refined and rounded edges. Materials: Metal.

Bench

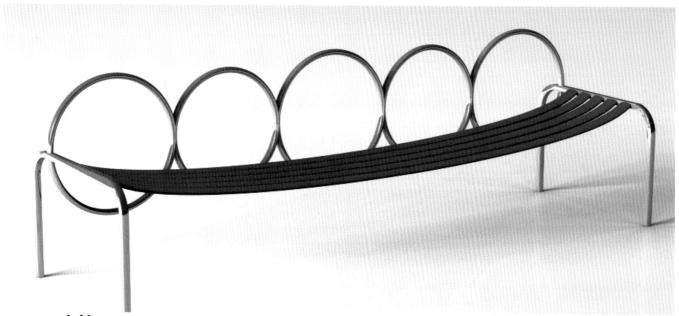

 Wheels 座椅
Wheels Bench

Wheels 座椅的设计灵感来自于奥运五环和吊床,是能给使用者带来快乐与舒适的趣味产品。尽管椅面看起来充满弹性,实际上它是由纯金属打造的。
制作方法:钢管 + 混凝土底座
材料:金属、混凝土

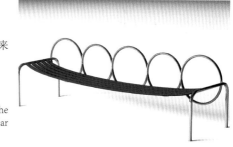

Wheels Bench is inspired by the Olympic logo and hammocks, and it is a fun and comfortable design that brings joy to the users. Even though the seating panel looks elastic, it is actually metal. How to Produce: All the parts are realized by tubular stainless steel. A concrete base is required to immobilize the object, as the design is very light. Materials: Metal, Concrete.

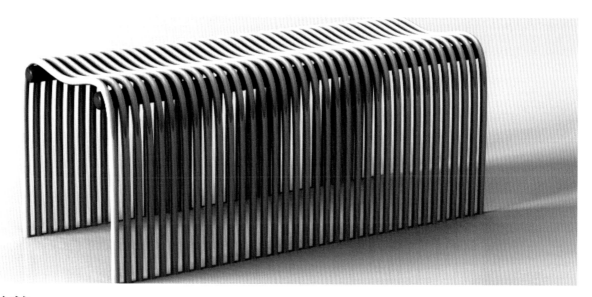

 Myriapoda 座椅
Myriapoda Bench

Myriapoda 多足类座椅拥有可爱的外形,其设计具有强烈的现代感和实用性。该类座椅较常规户外座椅更小巧玲珑,可同时用于室内设计。
制作方法:用连接管连接所有折弯的钢管
材料:金属

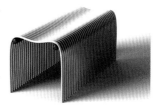

Myriapoda is a cute bench inspired by the Centipedes. The design is quite modern and functional, and it is smaller in size and could also be fitted into indoors. How to Produce: Stainless steel metal tubes are bended, and attached to the connector pipes. Materials: Metal.

座椅

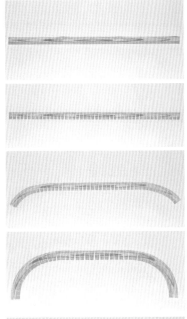

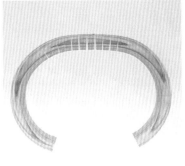

Shared Space III 座椅
Shared Space III Bench

设计：Chris Kabel　　Design: Chris Kabel
完成时间：2011　　　Year: 2011

对于喜欢亲密空间的人来说，Shared Space III 无疑是最佳的选择。该座椅用 10 米长的木梁加工成圆形，同时坐上三个人就可以形成一个小巧的私密空间，将所有的外部干扰摈弃在圈外，就像被大树拥抱住一样。

The circular bench is made from one 10-metre-long wooden beam. The bench really works. If you sit with three people or more in it, it automatically becomes a very intimate space where the outside world disappears. You really feel embraced by the tree.

挤压座椅
Extrusions Bench

挤压长椅由 6 张挤压的、镜面抛光处理过的铝制长椅组成。这款长椅并未配备任何的固定装置,而是用一整块铝块经过世界上最大的挤压机挤压而成,形成一张拥有椅腿、椅座和椅背的长椅。这款长椅原型最早的是一个长 100 米的户外装置。该项目从提出设计理念到建造成型共耗时 18 年,采用航空和航天工业技术制作,以建成世界上最大的、由金属制成的挤压长椅。同时,这个项目也是设计师 Thomas Heatherwick 展出的首个限量版作品。优雅的铝制长椅拥有独特动感的外形,挤压过程中形成的平行线构造出随意而粗糙的线条感。

The Extrusions include six extruded, mirror polished, aluminium benches made without fixtures or fittings, which have been produced by the world's largest extrusion machine. Heatherwick Studio commissioned a specially designed die through which aluminium was queezed into a chair profile, complete with legs, seat and back. The resulting exhibited extrusions are the early prototypes for a final outdoor installation, a 100-metre-long piece that tangles into an extraordinary form. The project, 18 years in the making, takes technology used in the aerospace industry to produce the world's largest ever extruded piece of metal. The project is also the first limited-edition work exhibited by Thomas Heatherwick. The graceful aluminium pieces each have a unique, dramatic form that combines the back, seat and legs into one dement. The sweeping parallel lines created through the extrusion process are contorted into random, gnarled endings: arbitrary swirling forms created through the inherent initiation and termination of the extrusion process.

座椅

新型长椅
Street Space Nrgzers Bench

| 项目档案 | Project Facts |

设计：Veronika Tzekova　　Design: Veronika Tzekova
项目地点：意大利　　　　　Location: Italy
完成时间：2010　　　　　　Year：2010

本案是一个新型的改进装置。原先的长椅尴尬地排列在街道边，正对着待租的商店门面。设计师采用了当地生产的黄色纱线将这些长椅连接起来。因而这些长椅成为一体，而且非常柔软，看起来就像是活力四射的光线。选择这些纱线是为了纪念这个地区纺织工业曾经的低迷时期。

The benches are an installation of strings. Conceived after an encounter with awkwardly placed benches on a street that had originally faced the facades of shop windows but now only look at the displays of "for rent" signs. The designer took locally produced yarn and extended the shape of the benches connecting them with strands of yellow. The resulting soft benches are like vivacious yellow rays of light and yet are not usable urban furniture. The use of the locally produced wool thread memorializes the social aspect of the economic decline of the textile industry in the region.

Bench

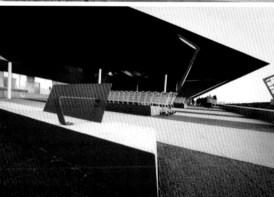

Costco in Melbourne, Australia
科思科大楼景观设计

项目档案	Project Facts
设计：NH Architecture	Design：NH Architecture
项目地点：澳大利亚，墨尔本	Location：Melbourne, Australia
完成时间：2011	Year：2011

本案的设计是为了提升科思科大楼整体的形象以及功能性。景观必须根据周围具体的环境而设计。在本案中，行人和车辆的关系处理得当，绿化做得也非常出色。

The proposed landscape design for the Costco building development provides areas that will contribute to the use of the building, the site as whole and which will encourage engagement with its surroundings. This will be achieved through the implementation of physical pedestrian and vehicular links and correspondence of vegetation types and planting themes.

座椅

More Bench Design

Bench

座椅

More Bench Design

Bench

座椅

More Bench Design

Bench

113

座椅

More Bench Design

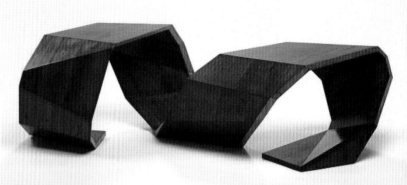

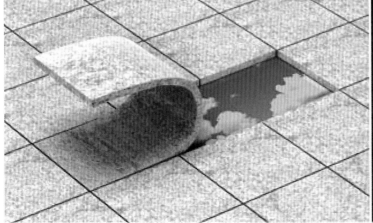

Bench

座椅

More Bench Design

Bench

座椅

More Bench Design

Bench

座椅

More Bench Design

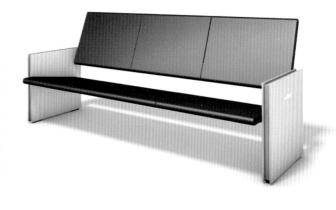
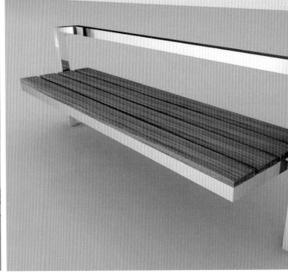

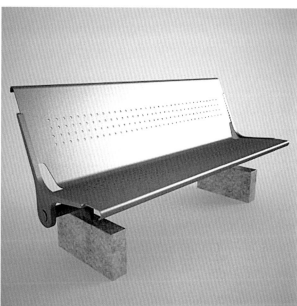

座椅

More Bench Design

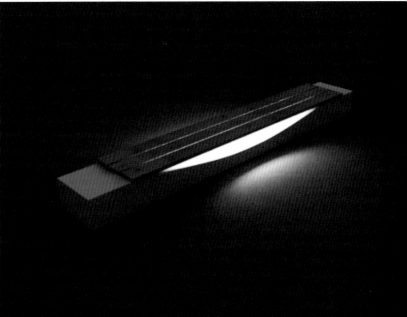

122

Bench

座椅

More Bench Design

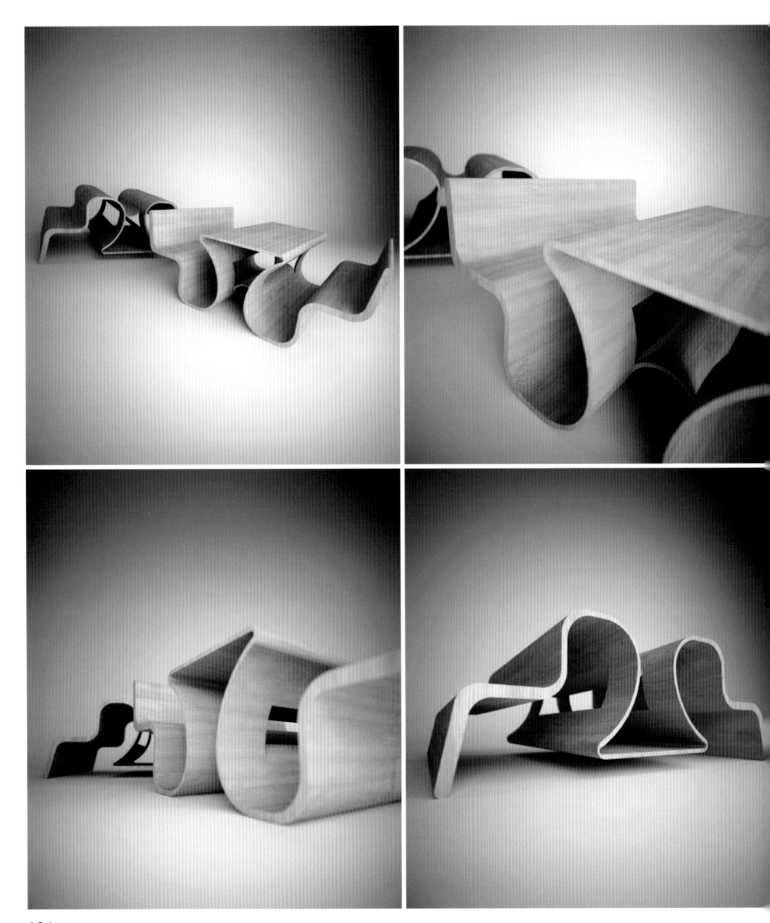

Bench

座椅

More Bench Design

Bench

座椅

More Bench Design

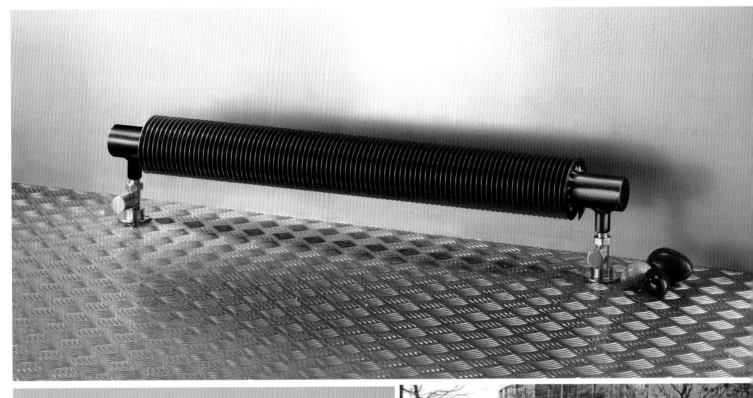

座椅

More Bench Design

Bench

座椅

More Bench Design

Bench

座椅

More Bench Design

Bench

座椅

More Bench Design

Bench

座椅

More Bench Design

Bench

座椅

More Bench Design

Bench

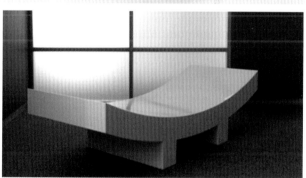

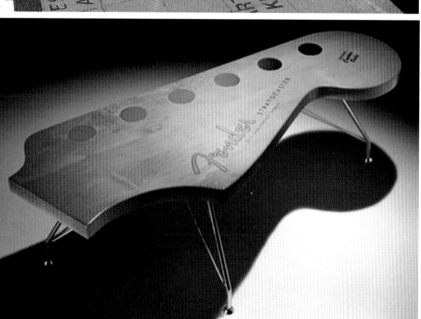

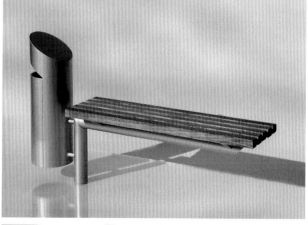

座椅

More Bench Design

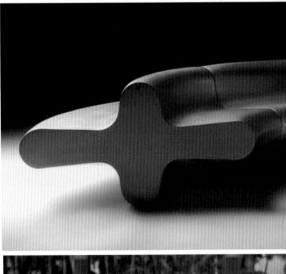

Bench

More Bench Design

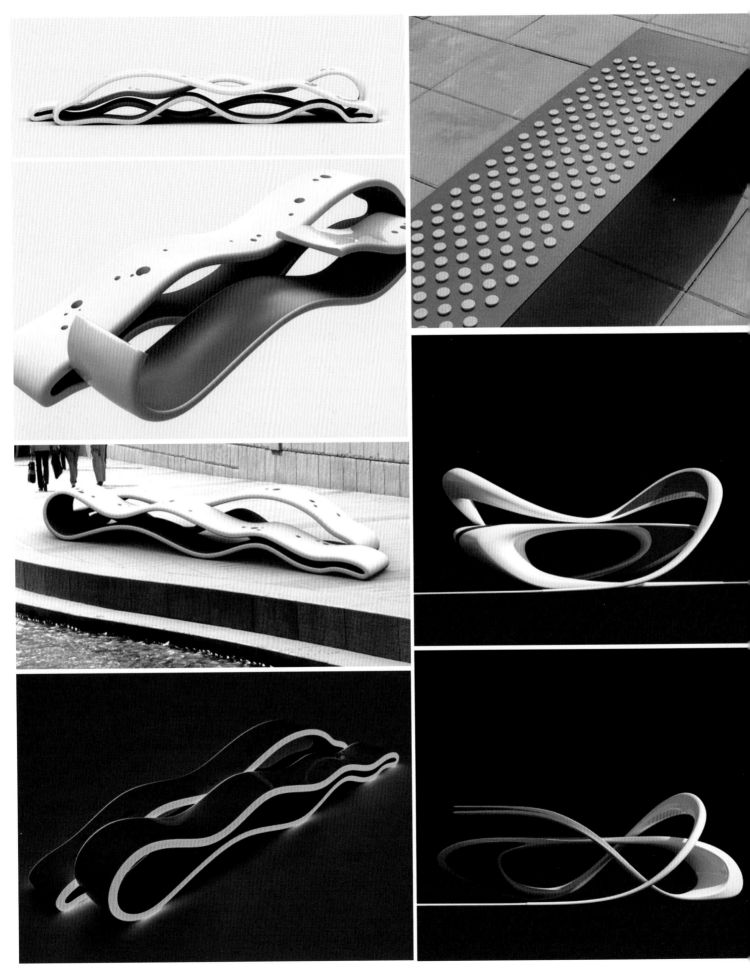

Bench

座椅

More Bench Design

Bench

座椅

More Bench Design

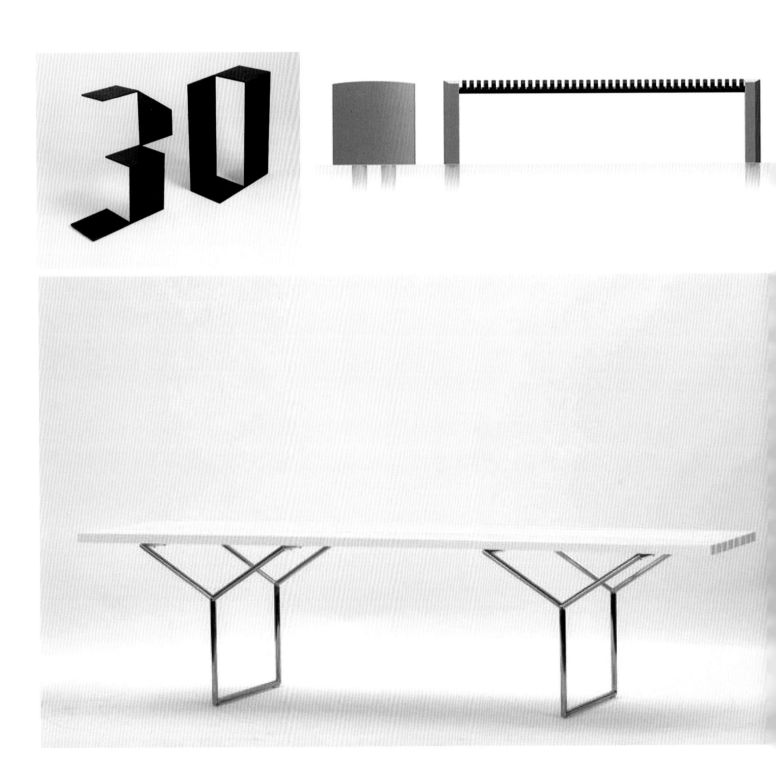

Bench

座椅

More Bench Design

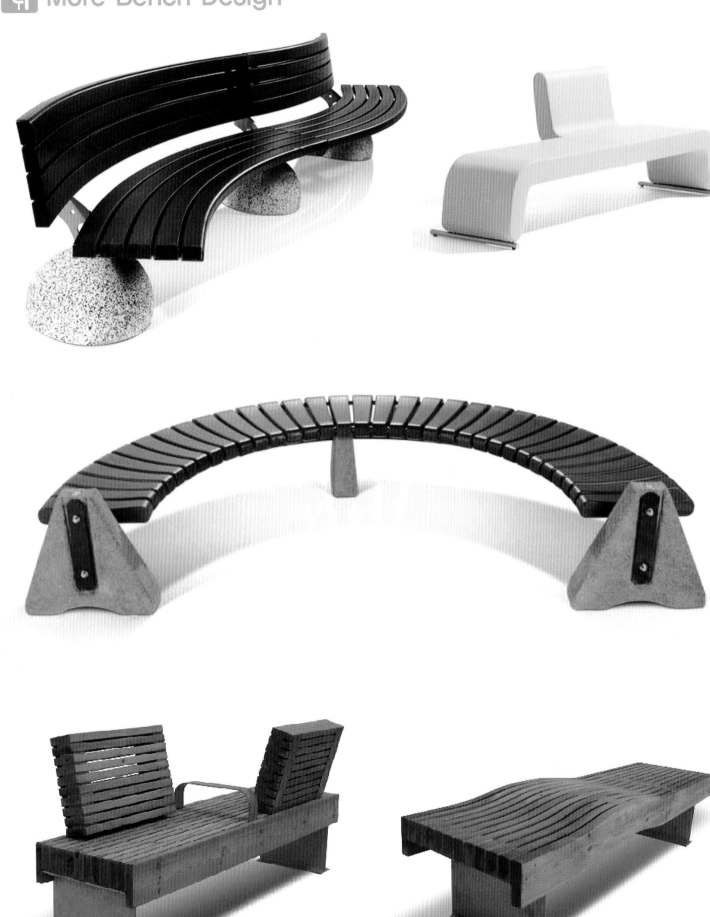

Bench

座椅

More Bench Design

Bench

座椅

Bench

座椅

Bench

座椅

More Bench Design

座椅

More Bench Design

Bench

座椅

More Bench Design

Bench

座椅

More Bench Design

Bench

165

座椅

 More Bench Design

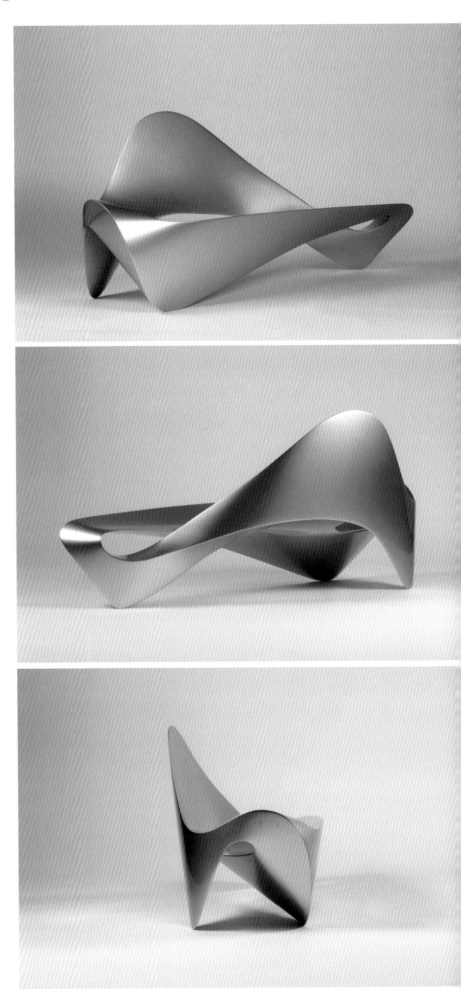

Bench

167

座椅

More Bench Design

Bench

自行车停放架
Bike Rack

自行车停放架

纽约市城市自行车停放架
City Bike Rack Design for New York City

项目档案　　　　　　　　　Project Facts

设计：Forms+Surfaces Designs　　Design：Forms+Surfaces Designs
项目地点：美国，加州　　　　　　Location：California, USA

受有机造型启发，"自行车公园"和其他自行车架不同，采用了一种独特的方式来应对日益突出的自行车停放和安全保管的问题。骑车者可以采用多种不同的方式来摆放自行车，同时提供了多个锁孔，用来锁住车杠和车架，增加自行车保管的安全性。整个自行车架由粗面防锈的不锈钢制成，主干部分可以采用表面装贴、现场浇筑等形式制作以实现灵活安排的最大化，或可事先组装布局。

Inspired by organic forms, the "Bike Garden" is unlike any other bike rack and provides a truly unique solution to the increasing challenges of bike parking and security. Its stems can be arranged in a wide variety of configurations to creatively accommodate almost any setting and provide rider with the added assurance of multiple locking points to secure the frame and wheels. Constructed entirely of rugged, corrosion-resistant stainless steel, Bike Garden's stems can be surface mounted or cast-in-place and may be purchased for maximum arrangement flexibility or in pre-configured layouts.

Bike Rack

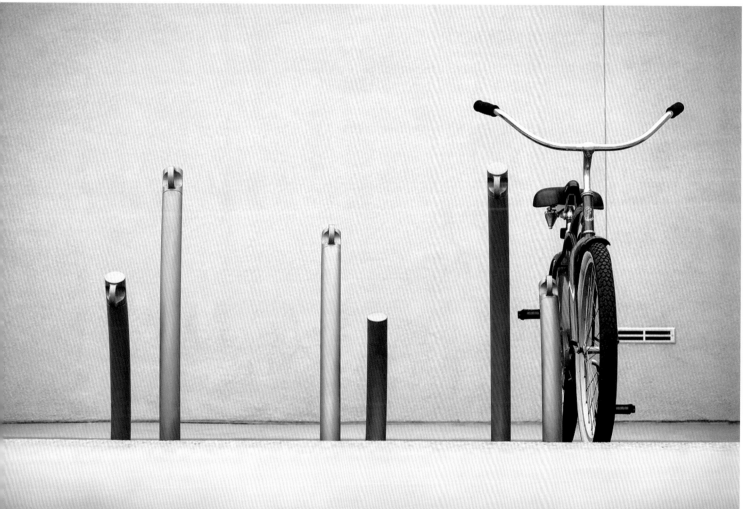

自行车停放架

港市自行车停放架
Bay City Bike Rack

项目档案

设计：Forms+Surfaces Designs
项目地点：美国，加州

Project Facts

Design：Forms+Surfaces Designs
Location：California, USA

本案简单且棱角分明的设计增添了多种成组的配置，不管是在视觉上还是在功能上对环境都有很好的影响。定向的车架拥有高效的空间布局，而灵活车架为人们提供了多角度的视觉效果，而且保障了高级别的安全要求。

The Bay City Bike Rack's simple, angled design allows for a wide variety of grouped configurations perpact for adding visual yet functional impact to any environment. Orienting racks in a single direction affords the most efficient space layout, while an alternating plan allows for multiple locking points when hight levels of security are required.

Bike Rack

弗里蒙特自行车停放架
Fremont Bike Rack

项目档案

设计：Forms+Surfaces Designs
项目地点：美国，加州

Project Facts

Design：Forms+Surfaces Designs
Location：California, USA

弗里蒙特自行车停放架光滑而耐用，是由耐腐蚀的铝以及橡胶缓冲器构成的。停放架不仅可以为自行车提供充足的锁扣空间，而且可以保证自行车框架以及轮胎的安全。

The Fremont Bike Rack is sleek, durable option with a wide stance that creates key locking points and the ability to fully secure the bike frame and both wheels. It is constructed of solid corrosion-resistant cast alurninum and an optional rubber bumper.

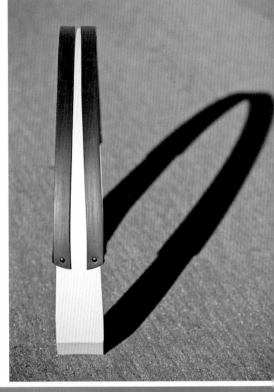

Bike Rack

自行车停放架

三角形自行车停放架
Trio Bike Rack

项目档案　　　　　　　　　　　Project Facts

设计：Forms+Surfaces Designs　　Design：Forms+Surfaces Designs
项目地点：美国，加州　　　　　　Location：California, USA

这个停放架是 Trio 生产线上的一个完美的补充。为了保持设计上的连贯性，这个停放架同样采用了三角形状以及夸张的中空设计。

The Trio Bike Rack is the perfect complement to our Trio product line. Providing an excellent opportunity for design continuity, the Trio Bike Rack draws on the same triangular shape and exaggerated void seen in both our Trio Bench and Trio Lighting.

Bike Rack

自行车停放架

国会大厦自行车停放架
Capitol Bike Rack

项目档案

设计：Forms+Surfaces Designs
项目地点：美国，加州

Project Facts

Design：Forms+Surfaces Designs
Location：California, USA

这种停放架稳固性很强，材料上选择耐腐蚀的铝；设计简单，而且节约空间，能融入各种环境中。

The rack's solid, corrosion-resistant cast aluminum body provides the strength necessary to stand up to continuous use while its simple, space-saving design allows it to engage with its surrounding environment as much or as little as desired.

Bike Rack

奥林匹亚自行车停放架
Olympia Bike Rack

项目档案

设计：Forms+Surfaces Designs
项目地点：美国，加州

Project Facts

Design：Forms+Surfaces Designs
Location：California, USA

这种停放架表面光滑，充满流线型的曲线之美。选用铝材料，较稳固，而且耐腐蚀性强；所占用空间较少，其他配置的选择就很多。尤其适合公园、校园等场合。

The Olympia Bike Rack's smooth, fluid curves combined with solid corrosion-resistant cast and aluminum construction make this rack a perfect choice for parks, corporate campuses and more. Its strand-alone space-saving design allows for an unlimited number of configuration options for ultimate design flexibility.

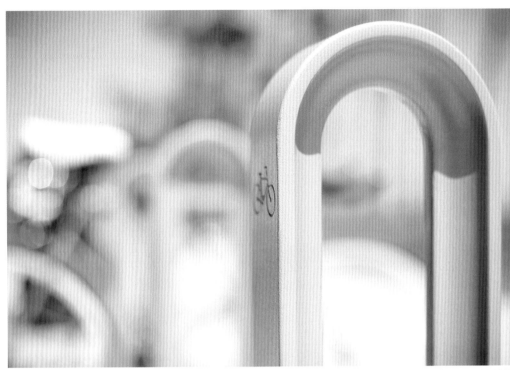

Bike Rack

自行车停放架

雕塑式自行车停放架
Sculptural Bike Rack

项目档案　　　　　　Project Facts

设计：Ralston & bau　　Design：Ralston & bau
项目地点：挪威　　　　Location：Norway

普通的停放架一般比较枯燥无趣，又不美观。所以，那些停放架都隐藏在棚屋和走廊里。这个停放架的设计解决了这些问题，而且还具有十足的美感。

Bike racks tend to be boring and unsightly, which is why they are often hidden in closets, sheds, and hallways. Problem solved! Sculptural Bike Rack is a reinterpretation of the rack for the aesthetically concerned.

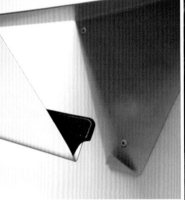

Bike Rack

 John Barbier 自行车停放架
John Barbier Bike Rack

项目档案	Project Facts
设计：John Barbier	Design：John Barbier
项目地点：美国萨凡纳大学	Location：Savannah College, USA

这个是设计师 John Barbier 的一个参赛作品。停放架的整体性和连贯性非常强，且注重线型美。相对于普通的停车架，其功能更加齐全。

This is a bike rack the designer John Barbier designed for a design competition. This design was selected as the winning entry. The designs focus on wholeness and continuity, and the beauty of line is very clear and impressing. Compared with other racks, they are of more functions.

自行车停放架

Ian Mahaffy 自行车停放架
Ian Mahaffy Bike Rack

项目档案

设计：Ian Mahaffy, Maarten De Greeve
项目地点：丹麦

Project Facts

Design：Ian Mahaffy, Maarten De Greeve
Location：Denmark

设计师们的目的就是设计一个稳固的、功能强大的、经济且安全的自行车停放架。整体的设计语言与人行道保持一致，并且易于人们发现和方便使用。最后的成品经过长期的测试和分析，在稳固性、生产成本以及安装的方便性方面都有充分的考虑。

The designer's goal was to create a robust, functional, economical and safe bike rack. The overall form of the new rack was kept in the same language as the existing sidewalk design to help the cyclist to easily and quickly identify the function. The development of the design went through a long testing and analysis phase with different design variants being weighed up against strength, production cost and ease of installation.

自行车停放架

 ## 多伦多自行车停放架
Toronto Bike Rack

项目档案	Project Facts
设计：Evi Hui, Olivier Mayrand, Michael Pham	Design：Evi Hui, Olivier Mayrand, Michael Pham
项目地点：加拿大	Location：Canada

这些停放架的设计呈现漫画风格，充满了快乐气息，让呆板的自行车都有了生气，更是让城市居民迸发出对自行车文化的骄傲与自豪感。设计的灵感来源于这条街本身文化与表现力的综合体现。同时也参考了流行的标识和象征符号。

The jovial comic-book style speech bubble designs celebrate and animate the stationary bike, while reflecting an emerging pride in the city's cycling culture. The inspiration came from Queen Street itself, and the street is a place of culture and expression. The designers also drew inspiration from pop culture signs and symbols.

Bike Rack

自行车停放架

 绿色自行车停放架
Green Bike Rack

项目档案	Project Facts
设计：Bike Arc's Designs	Design：Bike Arc's Designs
项目地点：美国	Location：USA

一个建筑公司设计了这个 U 型的自行车停放架。停放架从地面以曲线的方式构成一个 U 形，不仅节约了很多空间，而且上面的顶棚还可以遮阳挡雨。这些停放架经过特殊设计，任何人都无法偷走自行车轮胎。

This U-shaped bike rack is a simple curved unit that combines the space-saving quality of hanging systems with a much easier to maneuver roll-in-place design. The units can stand alone as the circular umbrella arc, incorporated into bus shelters or fully covered in the double-loaded tube design. These racks are also designed in a way that no one can steal your bike's wheels.

Bike Rack

 Holbech 自行车停放架
The Holbech Bike Rack

项目档案　　　　　　　Project Facts

设计：Holbech Design　　Design：Holbech Design
项目地点：丹麦　　　　　Location：Denmark

Holbech 设计公司在丹麦的第二大自行车城市欧登塞设计了这个单独的停放架。停放架不仅能够满足基本的自行车存放功能，而且在功能性和灵活性方面也有了很大的突破。材料上选取不锈钢，极光滑的表面，有着可爱的情趣。

Holbech Design, in Denmark's second cycle city Odense, developed this standalone rack. It is a fantastic new twist on providing secure, theftproof storage of bikes in public spaces. It is functional and flexible and made out of glass-blown, rust-free steel that has a lovely finish.

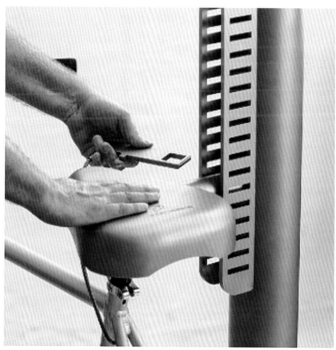

自行车停放架

B-Park
B-Park

项目档案	Project Facts
设计：Trust design	Design：Trust design
项目地点：美国	Location：USA

这种自行车停放架是用钢板制作的，设计最初是想放在达芬奇公园的。停放架大小基本上比较固定，颜色却是灵活多样的。停放架是独立的，避免自行车撞倒在一起。

"B-Park" is a bicycle rack designed from steel plate girders and bent for Vinci Park. Flexible as possible in color than the size of the rack, the separator plate seems to avoid that the bikes will collapse like dominoes.

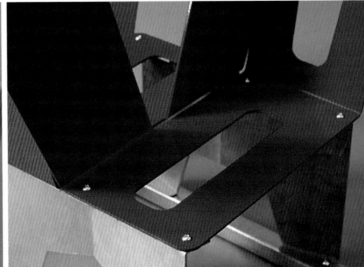

可丽耐座椅自行车停放架
Corian Bench Bike Rack

项目档案　　　　Project Facts

设计：Matt Gray　　Design：Matt Gray
项目地点：美国　　Location：USA

这是一个特别的设计，它既是长椅，也是自行车停放架。可丽耐人造大理石表面经过非常规的设计方法，构成了这个流动感十足的停放架。而且停放架在使用上和维护上都是可持续性的。

It is a combination bike rack and bench seating hybrid, and also the result of a challenge to integrate corian surfacing into an unconventional design solution. In addition to providing urban furniture that facilitates walking and bicycling, "bench rack" upholds in its production process of the US green building council's standards for sustainable operations and maintenance.

"郁金香奇趣"和"草"自行车停放架
Tulip Fun Fun and Grazz Bike Racks

项目档案　　　　　　　　**Project Facts**

设计：Design Studio Keha3　　Design：Design Studio Keha3
项目地点：爱沙尼亚　　　　　Location：Estonia

设计公司设计了两种自行车停放架——"郁金香奇趣"和"草"。在材料上选择的是具有灵活性的橡胶，这样，自行车的摆放位置、高度和形状的选择能够多样化。设计的灵感来源于郁金香和草地，不断重复的设计在大自然和城市之间筑起了一道人工屏障。这两种停放架可以通过楔形锚装置固定在地面上，也可以直接插入混凝土地面中。"草"停放架是由塑料包装的拉网钢丝绳构成的，上部分呈环形。

Design Studio Keha3 has created two bike racks——"Tulip Fun Fun" and "Grass". Made from flexible elastic, both systems accommodate multiple directions of bike placement as well as various heights and frame types, drawing from tulip and grass fields. The repeating forms create an artificial barrier that reinserts the versatility of nature into the urban environment. The racks can be affixed to the ground by using wedge anchors or by casting the components directly into concrete. "Grass" is made from plastic covered metal cables, with metal loops at their ends for attaching locks.

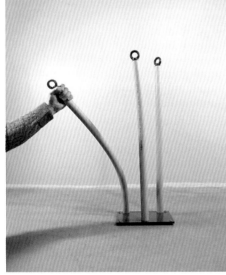

Bike Rack

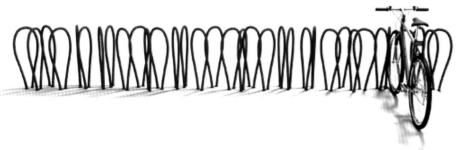

自行车停放架

巨型梳子自行车停放架
Gigantic Comb Bike Rack

项目档案 | **Project Facts**

设计：Helen Morgan
项目地点：美国佛吉尼亚州

Design：Helen Morgan
Location：Virginia, USA

这个巨大的手工制作的梳子状自行车停放架成为了佛吉尼亚州一道独特的风景。梳子停放架重达400磅，是通过船运到的。设计时参考了阴阳榫的结构原型，而梳子间的头发是用粉末镀层钢板构成的。

This gigantic handcrafted comb bike rack adds a fantastic dose of surreality to the urban fabric of Roanoke, Virginia. The 400 pound comb was shipped off to the city as part of a fun public art project. The rack used full mortise and tenon construction, while the hair is made from powder coated steel.

Bike Rack

小轿车式自行车停放架
Car Bike Rack

项目档案

Project Facts

设计：Cycle Hoop 公司
项目地点：伦敦

Design：Cycle Hoop Company
Location：London

这种停放架和汽车一般大小，可容纳十辆自行车，另外还设计了一个打气筒。在英国和部分欧洲城市都安装有这样的自行车停放架。设计师们还设计了一个自行车箍，用于保证自行车的安全。

The rack is the same size as a car and fits 10 bikes, and there's even a pump attached. They're installed around the UK and through some of Europe. The designers also designed a bike hoop used to turn existing street furniture into a safe and practical place to chain your bike.

More Bike Rack

Bike Rack

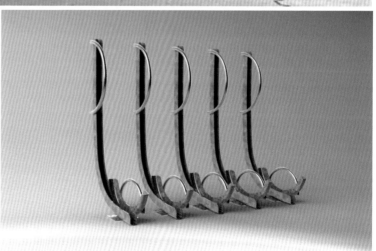

自行车停放架
More Bike Rack

Bike Rack

More Bike Rack

Bike Rack

自行车停放架

More Bike Rack

Bike Rack

205

自行车停放架

More Bike Rack

Bike Rack

自行车停放架

More Bike Rack

Bike Rack

候车亭
Bus Stop & Shelter

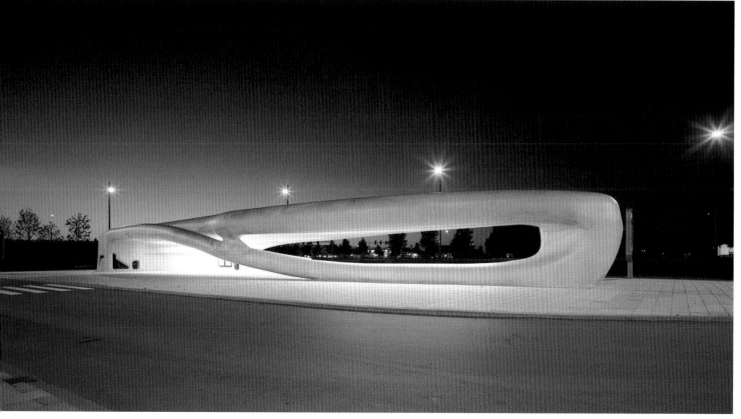

211

荷兰 Zuidtangent 专线上的候车亭
Bus Stop on Zuidtangent, Netherlands

项目档案	Project Facts
设计：From AtoB Public Design	Design：From AtoB Public Design
项目地点：荷兰，Zuidtangent	Location：Zuidtangent, Netherlands

Zuidtangent 地区的核心地带，即哈伦市与斯希普霍尔机场之间的地区，是世界上最繁忙的公共汽车线路之一。在阿姆斯特丹大都市圈，每天都有成千上万的人使用这个高效、便捷、舒适的公共交通系统。这一切都要归功于线路上红灰相间的候车亭，亮眼而结实的透明外壳设计与 Zuidtangent 居住区的现代化气息搭配得非常完美。

车站的设计必须满足几个条件，尤其是光照的需求。本案在照明方面相比传统的候车亭有个突出的优势就是方便维护。"Fortimo LED LLM 组件功率能达到 80%，并且能够使用 10 年。相比之下，传统的荧光灯管每两年就要更换一次。如果考虑到所有候车亭的照明需要的话，比如，路线信息和广告栏的照明。这样便可以节省大量的能源消耗。"Marcel Slop 还补充道："Fortimo LED LLM 的特别之处还在于，它有一个反射罩，能将光线向下折射，形成一个完美的圆锥形。除此之外，灯具的小巧造型和低调的外形适用于任何无论规模大小、造型统一的候车亭。"

The core section of the Zuidtangent region, between Haarlem and Schiphol Airport is one of the busiest dedicated bus lanes in the world. Every day, tens of thousands of people in the Amsterdam metropolitan area travel on this high-quality, quick and comfortable public transport system. And this is partly thanks to the new bright red-and-grey bus stops. Unlike the original waiting areas, the new stops provide excellent shelter. The stylish yet sturdy transparent shell design perfectly matches the modern feel of Zuidtangent's residential areas.

The design had to satisfy several conditions, particularly for lighting. One significant advantage over the old bus stops in terms of lighting is ease of maintenance. "The Fortimo LED LLM module bums at 80% of its capacity and is expected to last around 10 years. By comparison, an average fluorescent tube is replaced preventively every two years. You also save significantly more energy if you consider all the bus stop lighting, including the travel information and the advertising column." Marcel Slop adds: "The great thing about the Fortimo LED LLM is that the luminaire has a reflector which directs the light downwards in a well-defined conical shape. In addition, the luminaire's compact form and inconspicuous presence make for a uniform appearance regardless of the size of the bus stop.

Bus Stop & Shelter

候车亭

巴西古里提巴候车亭
Bus Stop on Curitiba, Brasil

项目档案

设计：Joseph Goodman, Melissa Laube, Judith Schwenk
项目地点：巴西，巴拉那省

Project Facts

Design：Joseph Goodman, Melissa Laube, Judith Schwenk
Location：Parana, Brasil

巴西古里提巴的公交系统是快速公交系统的典范，在活跃城市方面起着很重要的作用。公交车的运行班次达到 90 秒钟一班，非常准点可靠。公交站的设计也很方便、实用、舒适，并极具吸引力。因此，古里提巴拥有世界上使用率最高、然而造价却最低的交通系统。这套交通系统有着许多类似地铁的特点：公交车畅通无阻，不受交通灯和交通阻塞的影响；上车前先收费，乘客快速上下车；不同的是，这些都是在地面进行的。大约 70% 的人选择使用快速公交系统出行，因此在这个 220 万人居住的大古里提巴市内，消除了交通堵塞，而空气污染也减轻了。

The bus system of Curitiba, Brazil, exemplifies a model Bus Rapid Transit (BRT) system, and plays a large part in making this a livable city. The buses run frequently——some as often as every 90 seconds and reliably, and the stations are convenient, well-designed, comfortable, and attractive. Consequently, Curitiba has one of the most heavily used, yet low-cost, transit systems in the world, It offers many of the features of a subway system——vehicle movements unimpeded by traffic signals and congestion, fare collection prior to boarding, quick passenger loading and unloading——but it is above ground and visible. Around 70 percent of Curitiba's commuters use the BRT to travel to work, resulting in congestion-free streets and pollution-free air for the 2.2 million inhabitants of greater Curitiba.

Bus Stop & Shelter

候车亭

未来派联网候车亭
Futuristic Networked Bus Stop

项目档案

设计：麻省理工学院 Senseable City 实验室
项目地点：美国

Project Facts

Design：The Senseable City Lab of MIT
Location：USA

EyeStop 是用于车站的概念性装置，可以让乘客在候车亭不再只是无聊地等候公交车的到来，等车的同时还可以进行很多其他活动。该装置由麻省理工大学的 The Senseable City Lab 设计，由太阳能供电，配有触摸展示屏，显示各种有用的信息，比如：公交车班次、或者到达某个站点的最短路线等。甚至，你还可以上网、查看天气情况、查询你所要乘坐的班车的具体位置、和移动装置进行互动，并且将这个候车亭当作社区信息栏，发布通知和广告等。最重要的是，这个候车亭完全改变了传统车站的外观。
只需手指轻轻一点，用户就可以找到他们的目的地，系统会展示出最短的公交路线，并实时显示所有相关班车的精确位置。EyeStop 会发出不同的光亮，提示即将到站公交车的距离。路人也可以在电子播报板上发布广告和社区通告，将车站的功能提升到一个人们碰面和社区信息交换的层次。除了展示信息之外，车站同时还是一个环境感应观测点。通过太阳能自我供电，收集实时空气质量信息和城市环境信息。EyeStop 就是一个城市计算机的实验项目，就像一条信息带，贯穿着整个城市，像柱子一样站立在地面，或者是以小亭子的面貌出现。它感受着有关环境的信息，同时又以一种可传播的形式传递给全市居民。

Bus Stop & Shelter

EyeStop is a concept device for bus stops which will allow the passengers to do a lot more activities than just sitting idle and wait. The Senseable City Lab of MIT has designed this solar-powered booth that is outfitted with touchscreen displays that will show necessary information like bus schedules or the shortest route for a certain destination. Even more, you will be able to browse the web, check the air quality, see the exact location of your desired bus, interact with a mobile device and use the booth as a community message board to post announcements and ads. On top of that, this booth will surely change the appearance of a traditional bus stop.

At the touch of a finger users can indicate their desired destination; the system will then display the shortest bus route from where they are and the position of all relevant buses real time. The EyeStop will glow at different levels of intensity to signal the distance of an approaching bus. Riders and passers-by can also post ads and community announcements to an electronic bulletin board placed on the bus stop, enhancing its functionality as a public space, a place to gather and exchange community relevant information.

In addition to displaying information, the bus stop also acts as an active environmental sensing node, powering itself through sunlight and collecting real time information about air quality and the urban environment. EyeStop is an experiment into urban computing: it may be considered an info tape that snakes through the city, rising up like a pole or cropping out of the sidewalk like a shelter. It senses information about the environment and distributes it in a form accessible to all citizens.

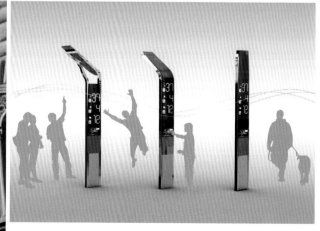

未来派联网候车亭
Futuristic Networked Bus Stop

项目档案	Project Facts
设计：Grimshaw Architects	Design：Grimshaw Architects
项目地点：美国，纽约	Location：New York, USA

候车亭的设计是优雅的，在满足候车亭所需的所有技术要求之外，它还是高效利用材料和耐用性的典范。该候车亭有两个落脚点，一个全玻璃背板还有一块半悬空的顶棚。顶棚上有硅架支撑，以铝材勾边与钢化玻璃边界相连。钢化玻璃和有槽式设计可以减少玻璃破裂而导致松脱的风险。根据环境的需要，候车亭的设计要求是透明的。一来设计独特，二来可以和城市景观融为一体，将视觉干扰降到最低。

The bus shelter is an elegant design which provides all the technical requirements of a shelter while exemplifying efficient material use and durability. It has two ground fixings, and a fully glazed rear panel and cantilevered roof. The roof system uses structural silicon to attach an aluminium profile to each edge of the laminated glass. The laminated glass and slotted design is intended to reduce the risk of glass becoming loose when broken. Contextually, the design needed to be transparent; it is distinctive in design yet blends into the urban landscape with minimal visual and spatial intervention.

Bus Stop & Shelter

候车亭

 Cemusa 西班牙一线
Cemusa Spain Line 1

项目档案	Project Facts
设计：Grimshaw Industrial Design	Design：Grimshaw Industrial Design
项目地点：西班牙，马德里	Location：Madrid, Spain

西班牙制造商 CEMUSA 委托 Grimshaw Industrial Design 公司设计出一套新的街道附属设施，能够拥有充分的灵活性和可延展性，让当地机构通过这些附属设施突出当地特色。设计的概念是运用一套高标准化的配件，通过这些配件的组合来达到全套设施的灵活性和可延展性。只有在发生了交通事故和破坏时，才需要进行设施维护，也只需要更换关键的部件。这大大降低了成本，同时也节约了维护时间。

一号线汽车候车亭的顶棚由单块玻璃面板和铸造铝臂组成。顶棚的所有固定点都分布在一条形状对称的铝制横梁上，同时该横梁还支撑着一块双边/单边的顶棚。这种一体化方式特别适合解决定点安装的问题。而支撑着玻璃板的支撑臂则是由失蜡法或包模铸造法制造而成。

Spanish manufacturer CEMUSA commissioned Grimshaw Industrial Design to design a new range of street furniture that would be flexible enough to be customized, and allow specifying local authorities the opportunity to adapt the furniture to emphasize their regional identity. The concept is based on a set of highly engineered components that can be used in various combinations to achieve flexibility and continuity throughout the range. When maintenance is necessary due to accident or damage, only the affected component needs to be replaced. The minimum costs as well as the time make the shelter out of commission.

The roof for Line 1 bus shelter comprises individual glass panels and cast aluminium arms. A symmetrically extruded aluminium beam supplies all the fixing points to support either a double or single-sided roof and accommodates all services. This integrated solution is particularly responsive to site-specific installation problems. The arms, which support the bus shelter glazing, are manufactured using a lost wax, or investment casting process.

Bus Stop & Shelter

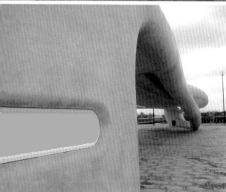

 霍夫多普汽车站
Hoofddorp Bus Station

项目档案	Project Facts
设计：NIO Architecten	Design：NIO Architecten
项目地点：荷兰，鹿特丹	Location：Rotterdam, Netherlands

本案位于一个广场中央，是一个岛状凸起的公共区域，是当地汽车运输服务的枢纽。一般这种建筑的设计都相对中规中矩，但本项目的目的在于创造一个强烈的个性形象，而非普通大众化。因此，该建筑设计为传统的奥斯卡·尼迈耶式，介于白色现代风和黑色巴洛克风之间。整座建筑完全由聚苯乙烯泡沫和涤纶制成，是世界上由人造材料所建成的最大结构 (50m x 10m x 5m)。

This project is located in the middle of a square and is a public area in the form of an island that serves as a junction for the local bus service. The design of this kind of building is generally neutral, but here the aim was to create a strong, individual image that was less austere and generic. Hence, the building was designed in the tradition of Oscar Niemeyer as a cross between white modernism and black Baroque. The building is completely made of polystyrene foam and polyester and is, as such, the world's largest structure in synthetic materials (50m x 10m x 5m).

候车亭

多伦多市汽车站
City of Toronto Bus Stop

项目档案
设计：Jeremy Kramer
项目地点：加拿大，多伦多

Project Facts
Design：Jeremy Kramer
Location：Toronto, Canada

这个全套设计是为多伦多首个20年街道附属设施合同而精选出来的。包括了交通候车亭、垃圾箱、长椅、公共洗手间、自行车架、信息栏、邮政和宣传栏等设施。

This comprehensive design was selected for the city of Toronto's first 20-year street furniture contract. The designs included transit shelters, waste receptacles, benches, public washrooms, bike racks, information columns, public posting columns and multi-publication structures.

Bus Stop & Shelter

223

La Dallman 汽车站
La Dallman Bus Stop

项目档案 / **Project Facts**

设计：La Dallman
项目地点：美国，芝加哥

Design：La Dallman
Location：Chicago, USA

车站位于一个由混凝土和石墙包围的平台上。混凝土结构和转折的石墙化身为长椅、挡土墙和支撑性结构。红木长椅设在连锁的混凝土结构和钢结构上，形成L形，满足行人多种需求，同时避开视觉障碍，可以清晰地看到远处行驶或靠近的汽车。大型钢框玻璃板负责阻挡大风，同时也形成了框景的视觉效果，和候车的乘客有直接身体接触的钢质元素被木材包裹起来。候车亭通过两边高、中间低的蝶形顶棚收集雨水，并将其引流到下方的混凝土蓄水池里。

The shelter is set within a platform defined by concrete and stone walls that are shaped and folded to serve as benches, retaining walls, and structural dements. Mahogany benches rest upon interlocking concrete and steel supports, forming an L-shaped plan that invites varied seating positions and protects users from the elements while allowing clear views to approaching buses. Large steel sash glass panels serve to block wind and frame views down the connecting Lift Station Path, and wood wraps the steel elements that come into direct contact with the occupants. The bus shelter collects rainwater through a butterfly roof, which drains into a cast concrete basin below.

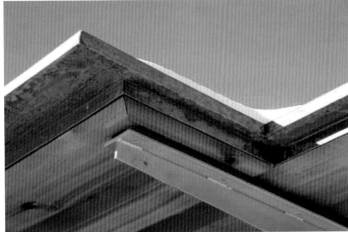

候车亭

Sean Godsell 建筑事务所设计的汽车站
Bus Stop——Sean Godsell Architects

项目档案 | Project Facts

设计：Sean Godsell Architects
项目地点：澳大利亚，墨尔本

Design：Sean Godsell Architects
Location：Melbourne, Australia

候车小屋主张人性化设计。当公共交通运行的时候，它是一个候车亭，而当公共汽车停驶的时候，它就变身成救急庇护站，常规的广告板可用作毛毯、食品和水的分发台，同时那块广告板也是一个小型的艺术展示空间。这个候车亭将来会实行太阳能供电，玻璃顶棚和后背板也将会被改装成数码投影屏。

The bus stop argues for compassionate infrastructure; it's a bus stop (when public transport is running) which converts into emergency overnight accommodation. The regular advertising hoarding is modified to act as a dispenser of blankets, food and water. As well the hoarding acts as a small gallery space where art can be exhibited and promoted. The stop has the potential to be solar powered and it is proposed that its glass roof and back double as a giant digital projection screen.

 安曼新候车亭
Amman's New Bus Shelter

项目档案	Project Facts
设计：Samar Hijjawi	Design：Samar Hijjawi
项目地点：约旦	Location：Jordan

候车亭的原型是令人印象相当深刻和稳固的。这个候车亭（或者其变形）都配有新式 Mupi 广告板、垃圾箱和长椅，在未来几个月内将会在全市范围内安装，是该市升级街道附属设施的一部分。

The prototype is pretty impressive and solid. This bus shelter (or variations of it) along with new Mupi advertising boards, trash cans and benches will be rolled out city-wide in the coming months, as part of the city's street furniture upgrade effort.

候车亭

候车亭
Bus Shelter

项目档案

设计：Pearce Brinkley Architecture
项目地点：美国，北卡罗来纳州

Project Facts

Design：Pearce Brinkley Architecture
Location：North Carolina, USA

候车亭的设计，简洁精致，建材包含了两种截然不同的元素：厚重的混凝土墙和座椅以及后期组装上去的金属面顶棚。无论是从立面图的角度还是从剖面图的角度看，混凝土墙和金属顶棚都构成了双 L 结构。顶棚采用了薄薄的聚碳酸酯材料，突出强调了轻盈和通透感。

混凝土墙和座椅或经过喷砂处理后覆盖上石板，或保留原始的状态，有着多样的功能性。所在校区的名字和地图可以直接铸造在混凝土墙面上。一片简单的木质椅面安装在墙体上，供人乘坐，候车亭顶棚结构则为等车的旅客遮挡太阳。透明的聚碳酸酯材料和薄薄的装饰玻璃安装在金属结构上，为旅客遮风挡雨。外层材料选择可透光材料，形成光影交错的景观。

The bus shelter is a simple yet refined architectural composition of two materially contrasting elements: a heavy cast concrete wall that serves as structure and as a bench, and a steel canopy frame, fabricated off-site, delivered by truck, and set into place. In elevation and section, the wall interlocks with the canopy forming a double "L" composition. The canopy skin, in this case laminated polycarbonate, further expresses the lightness and translucency of the canopy.

Sand blasted, clad with slate panels, or left in a natural state, the cast concrete wall has a versatile materiality. The name and map of each campus can be cast or applied directly to the concrete wall. A simple wooden bench fabricated of Ipe is attached to the wall providing a comfortable place to sit. The canopy structure and its associated skin provide shade and shelter. Either translucent polycarbonate panels or patterned laminated glass are attached to the steel frame providing a weather tight cover. The choice of skin further allows light to filter through animating the space with ever changing shadows and patterns of light.

Bus Stop & Shelter

候车亭

 Laing O'Rourke 候车亭
Laing O'Rourke Bus Shelter

项目档案	Project Facts
设计：Matt Emerson	Design：Matt Emerson
项目地点：澳大利亚，悉尼	Location：Sydney, Australia

这是一个长 72 米、宽 2.1 米的中转汽车站。设计方 Caldis Cook Group，TIDC，LOR 和 Adshel 与负责单位新南威尔士州铁路公司合作，令候车亭可以容纳相应的旅客及满足包括 160 勒克司的照明设施以及闭路电视系统等特定的要求。

The project is a 72-meter-long, 2.1-meter-wide bus interchange shelter. Working in association with the principal, Railcorp, their designer Caldis Cook Group and TIDC, LOR and Adshel were able to deliver a shelter designed to accommodate a number of client and project specific requirements including 160Lux lighting and OCTV cabling.

Bus Stop & Shelter

 威灵顿市候车亭
Wellington City Bus Shelter

项目档案   Project Facts

设计：Matt Emerson   Design：Matt Emerson
项目地点：澳大利亚，悉尼   Location：Sydney, Australia

Adshel Infrastructure 与威灵顿市议会紧密合作，为中央火车站和 Stout 街旁的人行道安装设施。设施总长超过 140 米，最长的一节，长达 22 米。该走道式候车亭连接火车站，为乘客们在风吹日晒的环境中提供了一个相对舒适的环境。

Adshel Infrastrure worked with the Wellington City Council for the supply and installation of walkways outside the Central Railway Station and along Stout Street. In total, over 140 metres of walkway were installed, the largest continuous section measured 22 metres. The walkways were designed to make the walk from the railway station to the offer pleasant for Wellington commuters in the frequently inclement weather experienced in the windy city.

候车亭

Euro Modul 候车亭
Euro Modul Bus Shelter

项目档案   Project Facts

设计：Euro Modul   Design：Euro Modul
项目地点：北欧地区   Location：Scandinavia

Euro Modul 公司是市区 / 街道附属设施的首选供应商。所有标准的候车亭都采用高标准和耐用的材料在制造。这些材料包括：不锈钢、铝、钢化玻璃、丙烯酸塑料和聚酯纤维。

Euro Modul is a leading supplier in east-south market place for street / urban furniture. All standard types of bus shelters are manufactured from the highest quality and durable materials: stainless steel, aluminium, tempered glass, the acrylic, polyester reinorced.

Bus Stop & Shelter

reddot design award
winner 2010

233

候车亭

 Montriol 候车亭
Montriol Bus Shelter

项目档案	Project Facts
设计：Design Montriol	Design：Design Montriol
项目地点：加拿大，蒙特利尔	Location：Montriol, Canada

候车亭设计从 STM 最新铸造完成的品牌标志"Mouvement collectif"获取灵感，由 Leblanc + Turcotte + Spooner 提出的设计方案中提供了一个成套而又灵活的解决方案。候车亭自带独立供电系统，其基座模型可以添加新的组件，形成不同大小、规模的组合，容纳不同数量的使用者。候车亭还设有一个互动栏，装有各种互动装置，其中包括动态数码显示屏和背光广告海报。内置太阳能光电系统保证候车亭的用电所需而无需接入电网。

Drawing inspiration from the STM's newly minted brand signature, "Mouvement collectif" the design proposal by Leblanc + Turcotte + Spooner offers a modular and scalable solution. Featuring a self-supporting structure, the concept enables the manufacturing of base models, with the possibility of joining several units together to create variable-size configurations that can accommodate larger or smaller numbers of users. The design features a communications column, which could house various components including dynamic digital displays and backlit advertising posters. An integrated solar power system will ensure lighting of shelters that cannot be connected to the power grid.

Bus Stop & Shelter

候车亭

Adshel 广告展示候车亭
Adshel Advertising Display Bus Shelter

项目档案	Project Facts
设计：Stoppress Design	Design：Stoppress Design
项目地点：新西兰	Location：New Zealand

Adshel 公司的户外创新运动仍在继续，旗下的新式 LED 广告展示技术被陆续公开。这些技术将会被运用在精选的 Adshel 所设计建造的候车亭中。而沃达丰公司最近率先在此种广告展示候车亭中投放有关移动上网的宣传广告。

Adshel has continued its recent streak of outdoor innovations with the release of its new LED advertising display technology, which will feature on selected Adshel-created bus shelters. And Vodafone's latest campaign to promote the joys of mobile internet is the first to put it to use.

Bus Stop & Shelter

Coleman 候车亭
Coleman Bus Shelter

项目档案	Project Facts
设计：Scott Freeman	Design：Scott Freeman
项目地点：澳大利亚，阿德莱德	Location：Adelaide, Australia

Adshel Infrastructure 公司和南澳政府合作的团队设计了阿德莱德威廉街上新有轨电车延长线上的 20 米长走廊式候车亭。走廊的中央顶梁是悬空式顶棚的视觉焦点，同时也安装了太阳能光照设施和标志，为使用者提供一个舒适的环境。顶棚结构的玻璃面板中内嵌了成套的太阳能电池板。

Adshel Infrastructure worked in partnership with the South Australian Government and their design team to deliver 20-meter-walkway shelters along the new tram extension in William Street, Adelaide. The walkway's central beam acts as a visual anchor for the cantilevered roof struts as well as housing solar powered lighting and signage, all of which serve to provide a comfortable environment for users. The roof structure is created with modular solar panels integrated into the glass panel.

汉堡汽车站 / 候车亭
Bus Station & Shelter, Hamburg

项目档案

设计：Blunck + Morgen Architecture
项目地点：德国，汉堡

Project Facts

Design：Blunck + Morgen Architecture
Location：Hamburg, Germany

汉堡 Poppenbutttel 车站建筑通过一条中央人行道将车站、火车站还有 P+R 停车场连接起来。这是 Poppenbutttel 交通枢纽重建和扩建工程的第一部分。设计概念要求将漂浮的翼楼结构与透明屋顶这两者合为一体，设计的原则就是将车站打造成一座空中雕塑。

The construction of the Hamburg Poppenbutttel station connects the bus stations to the train station and the P+R parking garage through a central pedestrian bridge. It is a first component of the restructured and expanded transportation junction of Poppenbutttel. The design concept called for an ensemble of buildings with a floating wing and a transparent roof. A guiding principle was the idea of building a sculpture in the air.

Bus Stop & Shelter

候车亭

Bus Stop & Shelter

候车亭

 More Bus Stop & Shelter

Bus Stop & Shelter

候车亭

More Bus Stop & Shelter

Bus Stop & Shelter

More Bus Stop & Shelter

Bus Stop & Shelter

候车亭

More Bus Stop & Shelter

Bus Stop & Shelter

候车亭

🚌 More Bus Stop & Shelter

Bus Stop & Shelter

More Bus Stop & Shelter

Bus Stop & Shelter

More Bus Stop & Shelter

Bus Stop & Shelter

照明设施
Street Lighting

海滩之灯
Licht am Strand

项目档案	Project Facts
设计：Museo Guggenheim	Design: Museo Guggenheim
项目地点：西班牙，拉斯帕尔马斯	Location：Las Palmas, Spain
完成时间：2010	Year：2010

本案中，灯的设计采用传统的斜挂大三角帆形状。它们取代了原有的28个相当耗电的灯柱。斜挂大三角帆形状灯的最高处为30米，因此在维修的时候就需要用到消防队的起重机。而下面的灯在维修的时候用普通的起重机就可以了。灯的主要材料为金属陶瓷，保证了灯光原来的颜色。另外，通过减少灯与灯之间的间距保证了照明的均匀。本案严格避免灯光污染，将直射光对周围环境的影响减少到最小。

The illumination point for the beach was the lateen sail design. These lighting fixtures replaced the former 28 light poles, which were heavy energy consumers. In addition, their height was about 30 meters. A special fire brigade crane is essential for their maintenance. The "lateen sails" with a height of 12-16m are however easily accessible and can use the normal maintenance of public lighting service cranes. Thanks to the ceramic metal halide discharge lamps, the color reproduction is improved. The new lights also ensure a more uniform illumination, which is achieved by reducing the distance between lights. In this design, it was extremely important to meet current standards for light pollution and to minimize the effects of direct light emission on the upper hemisphere (the sky).

Street Lighting

赫尔本博物馆的灯
Field of Light at the Holburne Museum

项目档案	Project Facts
设计：Bruce Munro	Design: Bruce Munro
项目地点：英格兰	Location: England
完成时间：2011	Year: 2011

本案可以称之为"灯地",在赫尔本博物馆展出。这一片"灯地"环绕着建筑,颜色与光线相互交错,异常美丽。5000多个灯泡由光纤连接构成网络状。白天的时候,"灯地"处于休眠状态;当夜幕来临的时候,这片灯地像花肆意绽放,充满了活力。

A "Field of Light" installation is on exhibition at the Holburne Museum. The field encircles the building, infusing the exterior space with an expanse of color and light in the dark. The installation consists of over 5,000 bulbs planted throughout the grounds of the museum, and an intricate network of fiber optic cables connects each acrylic stem crowned by a frosted sphere to the collective sculptural piece. "Field of Light" remains dormant during the day, but when night comes, the sculpture blossoms into a glowing, vibrant expanse.

照明设施

 街头灯管
Street Lighttube

项目档案 | Project Facts

设计：Marco Hemmerling
项目地点：瑞典，日内瓦
完成时间：2010

Design: Marco Hemmerling
Location: Geneva, Switzerland
Year: 2010

本案的设计概念源于空间认知的两个方面。第一方面是扩大中心柱和周围树木之间的关系。通过加强树木与广场中心的联系，将广场上垂直方向的元素聚拢起来。第二方面是在联系中寻找个性和特色，打破传统。抽象的想法转变成具体的形式，使得树木、中心柱还有广场都融入到一个颜色渐变的灯结构中。

The concept behind the lighting installation is based on two aspects of spatial perception. The first is to amplify the relationship between the structure's central column and its surrounding trees, by reinforcing the correlation of the trees with the centre of the square and connection of the vertical elements of the site. The second aspect of design is that though being a connector, the lighttube manages to establish its own presence, derived from the evolutionary principle of growth. The abstract idea transforms into a shape that puts the trees as well as the column and space into a new light, incorporating a constant change of color for the illuminated, trees and the membrane structure to support the idea of evolutionary transformation, while at the same time generating an ever-changing perception of the scenery.

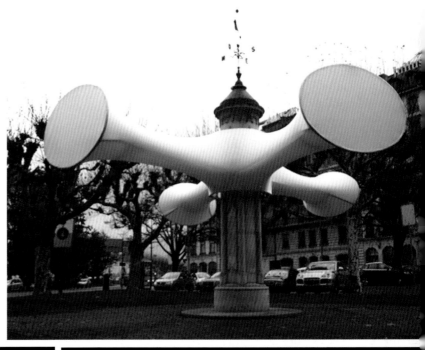

Street Lighting

1000 封诗信
1000 Poems by Mail

项目档案	Project Facts
设计：Luz Interruptus	Design: Luz Interruptus
项目地点：西班牙，马德里	Location: Madrid, Spain
完成时间：2010	Year: 2010

本案是为 2010 年西班牙诗节而设计的。1000 个白色的信封悬挂在公园里，每一个信封都有微光，并且装着一首 17 个诗人里面其中一位写的诗。因此，这里成为了诗歌朗诵表演的绝佳之地。

The installation was created for 2010 Spanish Poetry Festival. 1,000 white envelopes were hung throughout the garden, each of which contained a small light and a poem specifically written for the event by one of 17 poets. For three days, the filled envelopes illuminated the space, which was also the site of poetry readings and performances.

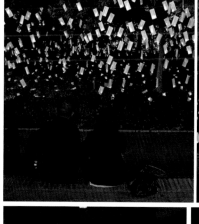

照明设施

字母路灯
The Street Alphabet Lamp

项目档案　　　　　　　Project Facts

设计：JDS Architects via　　Design: JDS Architects via
项目地点：丹麦，挪威　　　Location：Denmark，Norway
完成时间：2011　　　　　　Year：2011

字母灯的设计理念突出了一种创新方法，这种方法通过在保持核心设计的前提下，融入特定的文本结构使不同地区的灯柱产生了差异性。灯柱上的字母符号几乎是看不见的，只不过是简单地利用人工在上面钻了一排孔，光线可以从中通过，从而达到了独一无二的能见度。

The Alphabet Lamp concept features a way of differentiating lamp posts on different territories by incorporating specific textual configuration without changing the core design. The alphabetic representation is almost invisible, just an array of random hand drilled perforates on simple round post that let some light through, making unique visibilities for every lamp post.

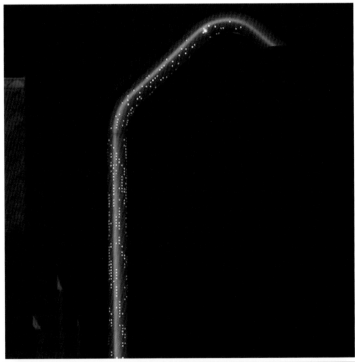

Street Lighting

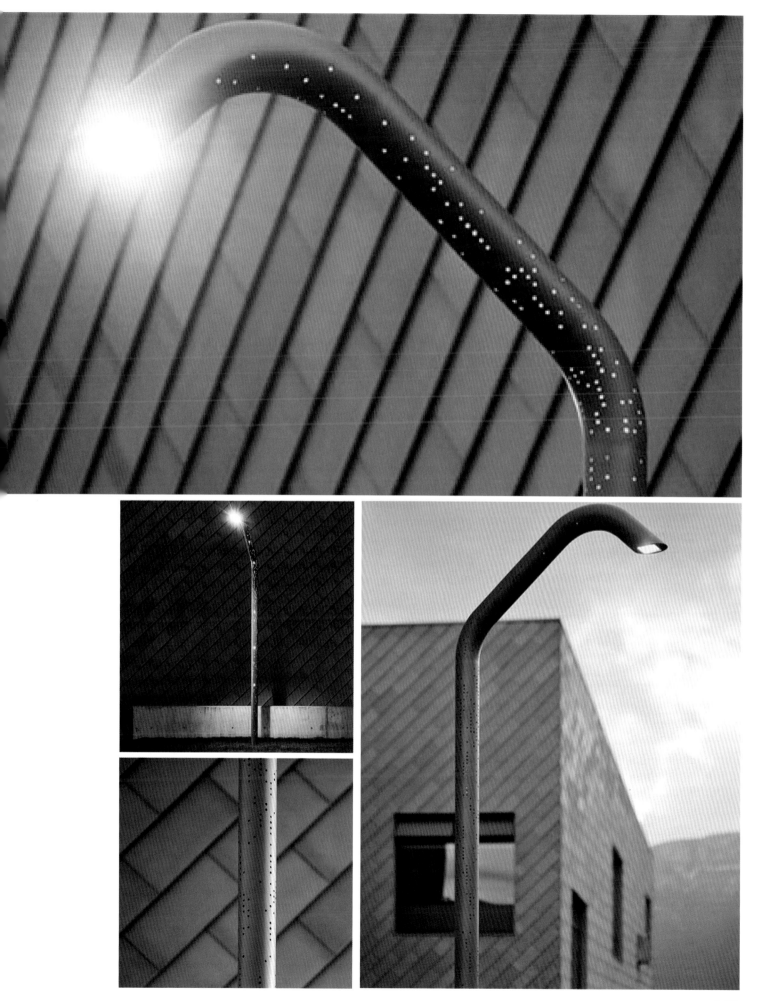

照明设施

Garscube Link
Garscube Link

项目档案　　　　　　　Project Facts

设计：7N Architects　　Design：7N Architects
项目地点：苏格兰　　　Location：Scotland
完成时间：2010　　　　Year：2010

Garscube Link 是一个公共区域，它将格拉斯哥北部和市中心联系起来，为行人和骑自行车的人打造了一个公共环境。设计师为 Garscube Link 项目取了一个非常别致的名字"凤凰之花"，这个名字参考了凤凰公园的名字，该公园在这里建造高速公路之前，就坐落于这片区域。"凤凰之花"区域由 50 个色彩斑斓的铝制"鲜花"设计而成，高 8 米，用独特的外观设计来吸引游客的注意，进而打破混凝土所带来的寂静的尴尬氛围。

Garscube Link is a new public realm intervention which re-connects North Glasgow back to the city centre for pedestrians and cyclists. The primary interface has been transformed from an inhospitable barrier to one that will be a positive threshold to the wider area. The new public realm is significantly wider than the previous underpass, held together by a single, flowing, red resin surface that doesn't constrain those using it to a single, confrontational, route. It is illuminated by a ribbon of 50 colored aluminium "flowers", fluttering through the space 8-meter up in the air, that draw the visitor through the route in deliberate contrast to the solidity of the concrete. The Garscube Link has been christened "The Phoenix Flowers", a reference to the former Phoenix Park which once occupied the site before the construction of the motorway.

Street Lighting

照明设施

Corvin Gate 街道照明
Corvin Gate Public Light

项目档案 　　　　　　　　Project Facts

设计：Zsolt Pyka 　　　　　Design: Zsolt Pyka
项目地点：匈牙利，布达佩斯 　Location: Budapest, Hungary
完成时间：2010 　　　　　　Year: 2010

这个 20 吨重的灯体结构使用柯尔顿耐腐蚀钢作为建筑材料，由两个非对称的灯脚及一道灯梁组成，高 4.8 米，长 35 米。其规模和结构使该物体成为一种建筑元素——门。这一 35 米长的落地式支座将代替正式的街灯扮演街道照明的角色。通过使用特殊的滤光器，它改善了无影灯的功能，能在灯脚的内侧构建发光二极管的像素，创造出绝妙的视觉效果。通过使用特别的专利技术能使 35 盏独特的 LED 无影灯一起发光。关于该灯体设计的一个重要参数是柯尔顿钢结构，它将街道照明与特殊的视觉效果结合在一起，这是对 LED 外墙的创新使用，媒体墙也采用了这种方法，之前达到这种规模的 LED 外墙还从未在街道照明设施中出现过。

Street Lighting

Made of corten steel, the 20-tonne lamp structure consists of two asymmetric feet and a beam positioned 4,8 meters high, stretching 35 meters long. This scale and structure makes the object an architectural element, namely a gate. In the future, the 35-meter-long console will take on the task of a public light, instead of the normal street lamps. Tungsram-schriders astral lamp has been improved by a special filter, providing an extraordinary visual effect that allows the pixel in the LED walls to be built in the inner sides of the feet of the lamp. The final optics are presented by the 35 uniquely planned astral LED lamps, using a special patent. The allocation on the two vertical LED walls is 10mm and on the vertical walls it is 20/10 mm. An important parameter of the object is that the corten structure is present as a construction connecting public lighting with special visual effects, which is an innovative use of the LED wall. Since this solution has mainly been used for media walls, a led surface of this size has never been built in a public lighting system before.

照明设施

3Rivers 行人交通灯
3Rivers Pedestrian Light

项目档案	Project Facts
设计：Forms+Surfaces Designs	Design: Forms+Surfaces Designs
项目地点：美国，加利福尼亚	Location：California, USA
完成时间：2011	Year：2011

3Rivers 是一种间接的行人交通灯，其精美的设计显得戏剧化而又十分低调。灯柱的蛋形截面以及椭圆形反射镜使灯体结构呈现出一种简洁的流线型外观。整个灯体都是由防生锈的铝材制成，外观十分惹人注目且十分耐用。3Rivers 是一种非常棒的照明灯，可以用于人行步道、公园、购物中心、校园以及其他的公共场所。

3Rivers is an indirect pedestrian fixture whose beautiful design is both dramatic and understated. The oval cross-selection pole and oval-shaped refector give 3Rivers a clean and streamlined look that is in a wide range of architectural and landscape settings. All components are rustproof aluminum. Striking in form and exceptionally durable, 3Rivers is a great lighting option for walkways, parks, malls, campuses and other public spaces.

Street Lighting

米兰 2010 年国际 LED 嘉年华
Milan LED Light Festival 2010

项目档案	Project Facts
设计：Fabio Novembre	Design: Fabio Novembre
项目地点：意大利，米兰	Location：Milano, Italy
完成时间：2010	Year：2010

如果将整座城市当成一块画布，那灯光就是画家手里的工具。米兰 2010 年国际 LED 嘉年华有三件作品呈现出了时间、记忆以及现代作风的概念。意大利设计师 Fabio Novembre 的作品《昨天，今天和明天》利用 LED 照明灯塑造出衣服的形状，这些"衣服"悬挂在"晾衣绳"上，照亮了整条大街。这一高技术性的作品再现了长达一个世纪的古老宣言，实现了传统与现代的对话。悬挂的衣服像是一面面旗帜，象征着一个崭新的意大利，它既时尚、风情又充满了生活的乐趣。

Using light as their tools and the city as the canvas, three works in the 2010 Milan LED Light Festival engage with ideas of time, memory, and modernity. In "ieri, oggi, domain" (yesterday, today, tomorrow) by Italian designer Fabio Novembre, LEDs recreate the forms of clothes hanging on clotheslines, illuminating the street. A highly technical reproduction of a centuries-old practice, the work converses with the dialogue between tradition and modernity. Novembre likens the hanging forms to make a new Italy, made of fashion, style, and the joy of life.

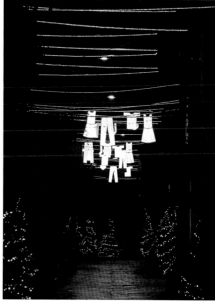

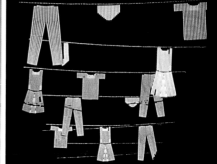

照明设施

行人灯柱
Light Column Pedestrian

项目档案

设计：Forms+Surfaces Designs
项目地点：美国，加利福尼亚
完成时间：2010

Project Facts

Design: Forms+Surfaces Designs
Location：California, USA
Year：2010

行人灯柱的建筑材料是不锈钢，利用 LED 线条灯或是线性荧光灯来照明，简单质朴，持久耐用，与周边环境轻松地融合在一起。灯柱的丙烯酸有机玻璃透镜可以在没有任何防护罩的情况下实现对称照明，也可以利用 180 度的不锈钢穿孔防护罩，此外还有定制的 108 度和 360 度防护罩。

Constructed of stainless steel and available with linear LED or linear fluorescent lamps, Light Column Pedestrian provided simplicity, ruggedness and easy integration into a wide range of exterior settings. The acrylic lens can be specified with no shield for symmetrical lighting or with a 108° perforated stainless steel shield. Custom 108° and 360° shields are also available.

Street Lighting

271

照明设施

Quartier Des Spectacles 照明规划
Quartier Des Spectacles Lighting Plan

项目档案
设计：Ruedi Baur，蒙特利尔建筑设计事务所
项目地点：加拿大，蒙特利尔
完成时间：2010

Project Facts
Design: Ruedi Baur, Montreal Architectural Design Studio
Location: Montreal, Canada
Year: 2010

该项目是蒙特利尔城市照明规划的一部分，探究了灯光创造的可能性，比如建立标识和表明身份。这是新近开展的试验性项目，通过将灯光投射到人行道上，创造出一种城市景观，成为蒙特利尔城市夜生活的一部分，将灯光设计和平面设计融合在一起。其照明系统给行人指明了如何去往附近的主要道路，当行人穿过十字路口时，利用灯光突出了他们的存在。其照明投射范围只限定在人行横道，清晰地显示出行人穿过街道的安全路径，同时向观光客展示了街区的文化活动。

The project is a part of the lighting plan in Montreal, and explores the possibilities of light for creating signage and expressing identity. This recent pilot project experiments with projecting light onto the pavement to mark the urban landscape. This intervention, realized as part of the Montreal all-nighter, brings light and graphic design together. The system shows pedestrians the way to major nearby venues, and highlights their presence as they cross the streets. The preliminary installation, still an experimental prototype, is made up of projectors suspended from towers and synchronized with the existing traffic lights at the intersection. The projections illuminate only the crosswalks, clearly indicating safe pedestrian passages across the street, which also shows visitors the cultural activities.

Street Lighting

照明设施

3XN 哥瑟姆路灯
3XN Gotham Street Lighting

项目档案	Project Facts
设计：3XN Design	Design: 3XN Design
项目地点：丹麦，哥本哈根	Location：Copenhagen, Denmark
完成时间：2010	Year：2010

3XN 设计了一种创新的 LED 灯具，利用一种特别的棱镜，成功地将太阳能运用到街道照明中。七盏 LED 路灯竖立在哥本哈根的贝拉中心前面，迎接即将到来的联合国气候变化大会。这种先进的技术使路灯释放出更多能量，比发光时使用的能量还要多，旨在实现路灯对二氧化碳的零排放。

3XN have designed an innovative LED luminaire, with the help of a specially developed prism, which uses solar energy for efficient street lighting. Seven lamps have just been erected at Bella center in Copenhagen in conjunction with the upcoming UN climate conference. The advanced technology resulted in the street lamps generating more energy than they use. The goal was to create a sculptural and CO_2 neutral street lighting solution.

Street Lighting

 ### B.Lux's 新品牌路灯
B.Lux's Brand New Street Light

项目档案	Project Facts
设计：GC Studio	Design: GC Studio
项目地点：西班牙，巴斯克	Location：Basque, Spain
完成时间：2010	Year：2010

Topa 是一种新型路灯，有统一鲜明的线条，由西班牙一家叫作 B.Lux's 的灯具公司所生产。这种路灯使用含有金属特性的棱镜，形成一个明亮的空心，向外投射出一束束灯光，这样光线就不会分散开来。该路灯由 GC Studio 设计，旨在制造出令人愉快的光影效果，并提升城市环境中的幸福感。路灯的建筑材料是镀锌钢材，表面涂有油漆，可以抵抗风吹日晒。此外，路灯的中性风格可以与各种现代建筑相搭配。

Topa is a new type of urban street lighting of which the monolithic, boasting sharp and architectural lines are manufactured by Spanish lighting company B.Lux. It's made of a metallic prism which creates a light-filled hollow and projects a beam of light outwards to create a pleasing effect of light and shade in street settings and enhance a sense of wellbeing in urban environments. Made of galvanised steel, although also available with a painted surface, it's extremely resistant to weathering. And its neutral style means that it suits all types of contemporary architecture.

骑士护柱
Knight Bollard

项目档案	Project Facts
设计：Forms+Surfaces Designs	Design: Forms+Surfaces Designs
项目地点：美国，加利福尼亚	Location：California, USA
完成时间：2010	Year：2010

骑士护柱的结构坚固耐用，与几何形状设计相结合，由粗壮的方形矮柱组成，在护柱四分之一的地方铸造了发光源，细节十分精美。这种护柱可以用作照明，也可以用于治安管理，保护建筑物和公共场所免受车辆干扰。

Knight Bollards combine rugged construction with a unique geometric design. The bollards consist of a robust square column with the light source positioned above a beautifully detailed quadrant casting. The bollards can be specified for lighting only or with an optional internal security core designed to protect buildings and public spaces against vehicle infringement.

Street Lighting

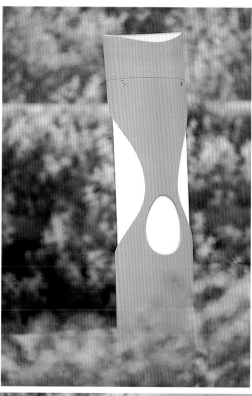

 轻型护柱
LightScale Bollard

项目档案

设计： Forms+Surfaces Designs
项目地点：美国，加利福尼亚
完成时间：2011

Project Facts

Design: Forms+Surfaces Designs
Location：California, USA
Year：2011

轻型护柱有流畅的有机线条，巧妙弯曲的轮廓，它可以与任何一种环境和谐相处。最初是作为一种路灯来设计，因其轻缓的弧形外观，这种护柱可以安装在街边，提供照明服务，并且不会占用步行空间。

With its smooth, organic lines and subtly curved profile, the LightScale Bollard is designed to settle harmoniously into any natural setting. Originally conceived as a path light, its gentle arc allows mounting beside a streets, providing light without taking up walk space.

照明设施

照明护柱
Light Column Bollard

项目档案
设计： Forms+Surfaces Designs
项目地点：美国，加利福尼亚
完成时间：2011

Project Facts
Design: Forms+Surfaces Designs
Location：California, USA
Year：2011

照明护柱的建筑材料是不锈钢，利用LED线条灯或是线性荧光灯来照明，简单质朴，持久耐用，与周边环境轻松地融合在一起。这种护柱可以保护建筑物和公共场所免受车辆的干扰。

Constructed of steinless steel and available with linear LED or linear fluorescent lamps, light column bollard provides simplicity, ruggedness and easy integration into a wide range of exterior settings, designed to protect buildings and public spaces against vehicle infringement.

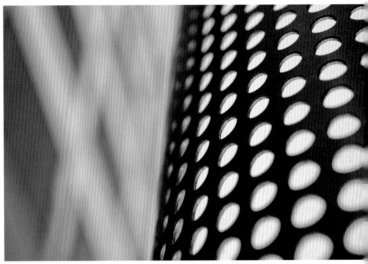

Street Lighting

照明设施

霞关广场翻新
Kasumigaseki Plaza Renewal

项目档案	Project Facts
设计：Thomas Balsley 联合公司	Landscape Architecture：Thomas Balsley Associates
项目地点：日本，东京	Location：Tokyo, Japan

这个项目位于东京中心，周围有政府机构和大量的私企。项目主要有两个大厦，一个是位于底层街道的榉树大厦，还有一个是位于高层街道的中心大厦。作为零售中心的入口，榉树大厦包括了大量的榉树，一个简单的木制甲板，公共座椅区，还有一个天然石头做成的天然楔子，构成了一个雾状喷泉。大厦的边缘用灯做成的"壕沟"，不仅是对古代壕沟的新的诠释，而且为大厦的边缘设计重新赋予一个新的定义。这个空间吸引了在这里工作的人和路人到此小憩。
中心大厦成为霞关广场的新入口，也是东京第一个高层建筑地标。一系列几何形状的景观带与建筑的几何形状相得益彰。咖啡馆和商店丰富了人们的活动区域。喷泉、路面铺设、植物和动态的灯光效果使得这个空间更加具有活力。

The Kasumigaseki building plazas are located in the heart of downtown Tokyo where government as well as major private business offices are concentrated. Due to large grade change, building entries occur at different but connected levels——the lower street level Zelkova plaza and the upper Central plaza.

Zelkova plaza at street level serves as an entrance to the retail portions of Kasumigaseki Building and Tokyo Club Building, consisting of mature zelkova trees, a simple circular timber deck, public seating, as well as a wedge of natural rock that forms a mist fountain. A light moat at the plaza edge is a 21st century reinterpretation of the historical moats that once defined this area and a light wedge wall gives a new dramatic definition to the plaza edge. The space attracts office workers as well as passers-by for a moment of relief from the surrounding urban context.

Central plaza serves as a new entrance plaza to the Kasumigaseki building, Tokyo's first high-rise and architectural landmark. A series of geometric landscape elements takes their shapes from the dynamic dialogue between building geometries. Cafes and shops keep the space fresh and alive with activity. Animated fountains, distinctive paving and planting patterns and dynamic lighting effects reinforce these settings as vibrant public spaces.

Street Lighting

照明设施

K3 一期
K3 - Phase 1

项目档案

设计：Urban Landscape Group
项目地点：匈牙利，布达佩斯
完成时间：2010（一期）

Project Facts

Landscape Architecture：Urban Landscape Group
Location：Budapest, Hungary
Year：2010 (phase 1)

K3项目一期是在原有的空间基础上，设计照明，结束空间功能上和形式上杂乱无章的现况，使之与周围的建筑和环境的特点相协调。

整个灯结构重达20吨，简单而具有功能性；灯结构由两个不对称的支脚构成，横梁有4.8米高，长度为35米。这种结构其实也是建筑的一部分，可以称之为"门"。35米长的横梁就为公共空间提供照明，现代的低消耗的LED灯得到了很多人的喜爱和欢迎。

另外更值得一提的是柯尔顿耐腐蚀钢结构。这种结构为空间营造了一种特别的灯光效果。这是对LED墙的一种创意性用法，通常这种用法被用在多媒体墙上。

The first element of 3K program is a special lamp for public lighting, which is finally named Corvin Gate Public Light. It puts an end to the disorder and chaos that is confusing functionality and formality in the square without interrupting the current characteristics of the surroundings.

The simple, functional, 20-ton lamp structure consists of two asymmetric feet and a beam positioned 4.8 meters high, stretching out 35 meters long. This scale and structure makes the object an architectural element, namely a gate. In the long run, this 35-meter-long console takes on the task of public light, instead of the normal street lamps. This modernized, low consumption, durable and economical LED lighting has appealed to all parties, receiving concordant support.

An important parameter of the object is that the corten structure is present as a construction connecting public lighting with special visual effects. This is an innovative use of the LED wall, since this solution has mainly been used for media walls.

Street Lighting

照明设施

Giulianova 纪念海滨
Lungomare Monumentale di Giulianova

项目档案

设计：360 景观设计建筑事务所
项目地点：意大利

Project Facts

Design：360 Landscape Architects
Location：Italy

这个项目重新诠释了 Giulianova 海滨中心地带。设计的想法是将其中一条交通道改建成一条人行道。这条人行道长 800 米，宽 55 米。整个空间连贯性很强，仍然注重和海的联系以及这个公共空间乃至整个城市的景观。新的空间艺术性很强，刺激了人们的参与欲望。

The project regards the redefinition of the central strip of Giulianova seafront. The idea of the plan is to reconvert to public-pedestrian use one of the two vehicular tracks organizing the traffic only on a side. The attempt is to rethink a band 800 meters long and 55 meters wide in which the spaces that follow each other are (from east to west) seaside. The plan is a landscape plan. The relation with the sea is searched also with one of its privileged points of view. New spaces activate the participation of people in the art landscape.

Street Lighting

照明设施

More Street Lighting

Street Lighting

照明设施

 More Street Lighting

Street Lighting

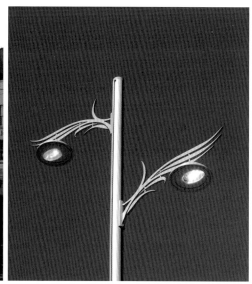

自动饮水器
Drinking Fountain

自动饮水器

DRINKMi 自动饮水器设计
DRINKMi Drinking Fountain Design

项目档案	Project Facts
设计：Aram Dikiciyan	Design: Aram Dikiciyan
项目地点：Milano, Italia	Location：意大利，米兰
完成时间：2011	Year：2011

DRINKMi 是一个公共的自动饮水器。采用"水瓶"这一形象的喝水标识，不仅因为公众熟悉，而且也为人们更正了自来水不是饮用水的错误观念。将水瓶旋转 90 度，水龙头就可以打开了。

"DRINKMi" is a public drinking fountain which, using the iconic image of a water bottle as an unequivocal drinking symbol, identify tap water that we often erroneously recognize as a second choice one as drinking water. The water faucet can be activated pitching the bottle by 90°.

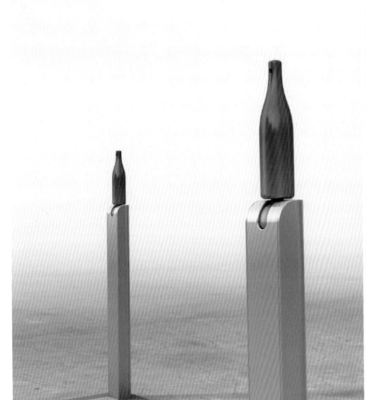

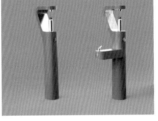

GlobalTap 自动饮水器设计
GlobalTap Drinking Fountain Design

项目档案	Project Facts
设计：IDEO	Design: IDEO
项目地点：美国，芝加哥	Location：Chicago, USA
完成时间：2011	Year：2011

这个自动饮水器是一个 5 英尺高的钢制结构，呈手肘状，有一个排水口。在使用的时候只需要按一下按钮，水就可以出来。自动饮水器的形状和使用原理意在鼓励那些愿意随身带水瓶的人们，提倡水瓶的循环利用以及减少塑料污染。目前饮水器采用的颜色是蓝色，目前橙色也在考虑范围之内。这两种颜色都很明显，在公园、湖边、公交车站、火车站都很合适。

This one was designed with a slim 5-foot-high steel stem, a thrusting elbow shape and a nozzle that directs a stream of water into the bottle when activated by a button. Its form and mechanism are meant to appeal to the new water-bottle-toting generation as a way to encourage bottle reuse and reduce plastic waste. The palette-blue is currently installed and orange is being contemplated. These two colors make the refilling station conspicuous in settings such as parks, lakefronts, bus stops and train stations, among the many urban sites under consideration.

Drinking Fountain

UBAN 自动饮水器设计
UBAN Drinking Fountain Design

项目档案 | Project Facts

设计：Forms+Surfaces Designs
项目地点：美国，加利福尼亚
完成时间：2011

Design: Forms+Surfaces Designs
Location: California, USA
Year: 2011

这种自动饮水器是由可循环利用的塑料构成的。除了抢眼的外观，它比用钢筋、混凝土和铁制的自动饮水器的重量轻很多。这种装置不仅能够提供干净的饮用水，而且还非常环保。

This kind of drinking fountain is made of 100% recyclable plastic. Aside from its eye catching aesthetics, it is also lightweight as compared to the traditional drinking fountains made of steel, concrete, and cast iron. With this product, we can now enjoy clean drinking water and consume it in an eco-friendly way.

自动饮水器

阿波罗 400 自动饮水器设计
Apollo 400 Drinking Fountain

项目档案	Project Facts
设计：Urban fountains+furniture	Design: Urban fountains+furniture
项目地点：美国，芝加哥	Location: Chicago, USA
完成时间：2011	Year: 2011

这种自动饮水器具有很强的耐用性和防破坏性，是阿波罗系列中专为残疾人设计的。总高度为 845 毫米，间隙高度为 680 毫米。

The Apollo 400 is a durable, vandal resistant outdoor disabled accessible fountain. It is just one of our fountains in the Apollo range designed for disabled access. Total height is 845mm, and clearance height is 680mm.

Drinking Fountain

自动饮水器

Hydro 300 自动饮水器设计
Hydro 300 Drinking Fountain

项目档案 | **Project Facts**

设计：Petr Hrusa　　Design: Petr Hrusa
项目地点：意大利，米兰　　Location：Milano, Italy
完成时间：2010　　Year：2010

在任何一个公园或城市广场都有一些不锈钢自动饮水器。本案的这种饮水器呈圆锥形，最上方有一个喷嘴。水经由这个喷嘴流出，流入地面的排水道里。

No well-equipped park or city square should lack a stainless steel drinking fountain. The jet from the more radically designed HD300 is installed on the top of a distinctive cone, from which the redundant water runs down to the drainage square around the fountain.

More Drinking Fountain

Drinking Fountain

垃圾桶
Litter Bins

垃圾桶

花托式垃圾分离箱
Dispatch Receptacle Litter Bins

项目档案

设计：Forms+Surfaces Designs
项目地点：美国，加利福尼亚
完成时间：2010

Project Facts

Design: Forms+Surfaces Designs
Location: California, USA
Year: 2010

这种垃圾箱设计非常独特，采用的材料坚固耐用，垃圾分类清理的设计也很合理。花托式的外形很是特别，垃圾箱是由铝皮做成的，旁边有一个铰链式的边门，方便人们处理垃圾。

The litter bins combine distinctive design, robust materials and versatile array of waste stream management options. Receptacles are made of heavy cast aluminum with a hinged side-access door for easy servicing.

Litter Bins

轨道形垃圾分离箱
Orbit Receptacle Litter Bins

项目档案

设计：Forms+Surfaces Designs
项目地点：美国，加利福尼亚
完成时间：2010

Project Facts
Design: Forms+Surfaces Designs
Location: California, USA
Year: 2010

这些垃圾桶设计多样，既可以用在室内，也可以用在户外。不锈钢和锻铁均是制作垃圾桶的最佳材料。锻铁不仅和不锈钢一样坚固，而且还具有颜色。单这一点来说，锻铁更具有优势。盖子是用铰链式的铝做成的。上面有形象的图像用以标明垃圾桶的各分类口。

The litter bins provide a wide array of design options, making them a unique solution suitable for indoors or outdoors. Orbit is available with a stainless steel or fused metal body. Fused metal adds a touch of color, while boasting the same durable characteristics of stainless steel. Both single-stream and split-stream models use hinged cast aluminum lids, with graphics clearly identifying the function of each opening.

Litter Bins

 三角形垃圾分离箱
Triad Receptacle Litter Bins

| 项目档案 | Project Facts |

设计：Forms+Surfaces Designs　　Design: Forms+Surfaces Designs
项目地点：美国，加利福尼亚　　Location：California, USA
完成时间：2010　　Year：2010

这些垃圾桶为公共空间的垃圾收集以及回收提供了更多的选择和面貌。可以单独使用，也可以组合起来使用，这样更具有特色和吸引力。

The litter bins offer a distinctive and versatile modular solution for public space litter collection and recycling. Receptacles can be used alone or may be arranged together in attractive geometric groupings.

运输型垃圾桶
Transit Litter Bins

项目档案

设计：Forms+Surfaces Designs
项目地点：美国，加利福尼亚
完成时间：2011

Project Facts

Design: Forms+Surfaces Designs
Location: California, USA
Year: 2011

这些垃圾桶的设计更具有吸引力，在垃圾收集与处理上更能体现一体化。材料上选择耐用、耐腐蚀的不锈钢。在空间上，有三个开口、三个独立的储物区，分别存放垃圾、纸张和饮料瓶。侧边开口也很大，方便人们扔垃圾。这些垃圾桶适合在机场、火车站、公交车站和大学校园内使用。

The Transit Litter & Receding Receptacle provides an attractive and efficient all-in-one solution for public space litter collection and recycling. Made from durable, corrosion-resistant stainless steel, Transit's space-saving design incorporates three separated openings and three independent internal compartments for conventional litter, paper and beverage containers. Large side access doors and a quick-action latch make Transit exceptionally easy to service. The bins are ideal for airports, train or bus terminals, university campuses and other environments.

Litter Bins

树木防护装置
Tree Protection

树木防护装置

尖沙咀树木保护
Tsim Sha Tsui Tree Protection

项目档案	Project Facts
设计：Studio Zhui	Design: Studio Zhui
项目地点：香港	Location: Hong Kong
完成时间：2011	Year: 2011

五个与众不同的灯柱热烈地欢迎着来访者，并且为居民提供着保护作用。其中的三个灯柱围绕着三棵树，另外两个灯柱分成两叉，看起来就像是人敞开双臂，邀请人们来到这里，尤其是老年人和小孩。

The five sets of distinctive lamp posts mean to represent the welcoming warmth to visitors, all-round protection and stem support for the district citizens. Three of them, each surrounding a decade-aged tree and the other two crotched, symbolizing the association's pair of opening arms, give the most caring and inviting gestures to the public, especially the young and the old.

Tree Protection

树木防护装置

国家原住民卓越中心
National Centre of Indigenous Excellence

项目档案	Project Facts
设计：Social Innovation Sydney	Design：Social Innovation Sydney
项目地点：澳大利亚	Location：Australia
面积：7 100 平方米	Site Area: 7,100 m²

这个项目是在原来的悉尼红坊区公共学校的基础上改建的，旨在创建一个多功能的住宅区域，融培训和教育为一体，为社区，郊区乃至整个城市服务。具有历史意义的建筑和宽阔的空间被重新设计，发展成五个主要的活动区域，每一个区域都有独特的功能。

这个学校除了有宽阔的教室和娱乐区域，同时也提供了可以容纳100人的住宿和食堂。原来学校西边的游乐场被改造成一个高水准的足球训练基地。一座三层楼高的多功能运动综合体被建成，包括一个室内运动大厅，多个活动室，一个长达25米的热水游泳池、更衣室、储物柜。

The National Centre of Indigenous Excellence is the redevelopment of the historic former Redfern Public School by the Indigenous Land Corporation as a multi-use residential, training and education facility catering for both the local community and rural and interstate groups.

The significant heritage buildings and vacant areas of the site have been integrated and developed for five major activities, each supporting the role of the centre.

The Eora Campus provides dormitory accommodation for visiting educational and sporting groups of up to 100 people, with associated sleeping, dining, classroom and recreational areas. A high-quality football training field, chiefly for use by the Eora Campus facility is located on the western playground of the former school, facing Cope Street. A purpose-built three level multi-use sporting complex, includes indoor sports halls and activity rooms, a heated 25m pool and associated change and storage areas.

Tree Protection

树木防护装置

More Tree Protection

Tree Protection

树木防护装置

🌲 More Tree Protection

Tree Protection

树木防护装置

More Tree Protection

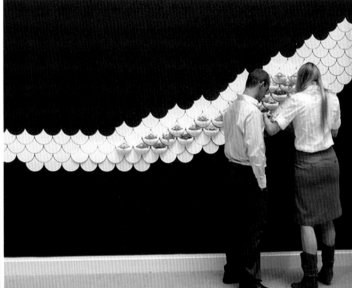

Tree Protection

图书在版编目（CIP）数据

公共景观集成.街景设施/广州市唐艺文化传播有限公司编著.——北京：中国林业出版社，2016.4

ISBN 978-7-5038-8440-5

Ⅰ.①公… Ⅱ.①广… Ⅲ.①景观设计－图集②城市道路－景观设计－图集 Ⅳ.①TU986.2-64 ②TU984.11-64

中国版本图书馆CIP数据核字(2016)第050816号

公共景观集成　街景设施

编　　著	广州市唐艺文化传播有限公司
责 任 编 辑	纪　亮　王思源
策 划 编 辑	高雪梅
文 字 编 辑	高雪梅
装 帧 设 计	杨丽冰

出 版 发 行	中国林业出版社
出版社地址	北京西城区德内大街刘海胡同7号，邮编：100009
出版社网址	http://lycb.forestry.gov.cn/
经　　销	全国新华书店
印　　刷	深圳市汇亿丰印刷科技有限公司
开　　本	220 mm×300 mm
印　　张	19.875
版　　次	2016年8月第1版
印　　次	2016年8月第1次印刷
标 准 书 号	ISBN 978-7-5038-8440-5
定　　价	318.00元（精）

图书如有印装质量问题，可随时向印刷厂调换（电话：0755-82413509）。